Schriftenreihe der Professur für Molekulare Lebensmitteltechnologie

Band 11

Institut für Ernährungs- und Lebensmittelwissenschaften (IEL)

# Lactic acid fermentation for the production of pyranoanthocyanins as food colorants

Dissertation

zur Erlangung des Grades

Doktorin der Ernährungs- und Lebensmittelwissenschaften

(Dr. troph.)

der Landwirtschaftlichen Fakultät

der Rheinischen Friedrich-Wilhelms-Universität Bonn

vorgelegt von

Elsa Anaheim Aguirre Santos

aus Durango, Mexiko

Bonn 2020

**Bibliografische Information der Deutschen Nationalbibliothek**
Die Deutsche Nationalbibliothek verzeichnet diese Publikation in der Deutschen Nationalbibliografie; detaillierte bibliographische Daten sind im Internet über http://dnb.d-nb.de abrufbar.
1. Aufl. - Göttingen: Cuvillier, 2021
Zugl.: Bonn, Univ., Diss., 2021

Angefertigt mit Genehmigung der Landwirtschaftlichen Fakultät der Rheinischen Friedrich-Wilhelms-Universität Bonn

Referent: Professor Dr. rer. nat. Andreas Schieber
IEL – Molekulare Lebensmitteltechnologie, Universität Bonn

Korreferent: Professor Dr. rer. nat. Matthias Wüst
IEL – Bioanalytik (Lebensmittelchemie), Universität Bonn

Tag der mündlichen Prüfung: 07.06.2021

Erscheinungsjahr: 2021


Nonnenstieg 8, 37075 Göttingen
Telefon: 0551-54724-0
Telefax: 0551-54724-21
www.cuvillier.de


1. Auflage, 2021
Gedruckt auf umweltfreundlichem, säurefreiem Papier aus nachhaltiger Forstwirtschaft.

ISBN 978-3-7369-7553-8
eISBN 978-3-7369-6553-9

## ACKNOWLEDGEMENTS

I would like to express my gratitude to Professor Andreas Schieber for all knowledge share and for taking me under his guide in this amazing Ph. D experience that has changed my life. Thank you for providing the financial resources to make the project possible.

I feel grateful to Fabian Weber “my Betreuer”, who help me during this journey, believing in this project and patiently motivating me in dark moments. It had always fun discussing our results, even trying to accept that our "Tierchen" will always do many things but not necessarily what we wanted. Thank you for sharing your research with me and the working group in Braunschweig.

I had such luck to work with so many beautiful people like Peter Heffels with whom sharing the office left me so many unforgettable memories, singing in the office, celebrating the first baby shower in the institute, playing with laboratory material and many others. Hannes Patzke, my first potential enemy that ended up being the nicest enemy I could find in Germany, I will never forget all the things you gave me your help with, and all my birthday cakes, definitely, Germans should adopt this tradition. My dear Sherlock! Michael Kaisers, you are a great person to share the office and our Friday of the playlist the best. I want to thank you for all those long days analyzing my samples, thank you very much for your help with the MS. Rita Capers-Weiffenbach , you play such an important role in the group, I would have not made it without all your help, ordering and organizing all the material. Your great heart and motivational words always kept me going. I have enjoyed our happy days in the laboratory and our days out a lot. Sandra Damm your disposition for helping others is admirable, thank you for giving us a nice smile and your kind words every day. Ulrike Schroeter, thank you for your help in all administrative aspects. I will never forget your coffee cup, you showed me how important it is to look for the things you want.

It is complicated for me to express all my gratitude to this great working group, it was amazing to work next to all of you. I will always remember the meetings every morning to drink coffee, discussing work inside or outside the office with all of you, my sincere thanks to all of you: Lena Larsen, Franziska Bührle, Sabrina Zimdars, Maria Linden, Lisa Abel, Sara Schmitt, Lara Etzbach, Ingrid Weilack, Timo Heinrichs, Nadine Schulze-Kaysers, and Michelle Feuereisen . Thanks to all of you I discover how great life is in this country, sharing traditions, food, teaching me your language, and sharing your research with me. I am very happy to consider all of you more than my colleagues, my friends.

Amin Lahnaoui I cannot thank you enough, for your patience, encouragement, thank you for your support during this thesis process and for being my rock in the toughest moments these years.

I want to thank my dearest friends Antonio and Carolina, there are no words to express what I feel about you two, you are life-changing persons and I am happy to call you my family. Elisa, Ana Eva, Fausto and Alejandro for all their support and encouragement, there was not a day I did not miss you.

Finally, I want to express all my love and admiration to my family, my father Manuel and my sisters, Denisse and Alyn but especially my mother Elsa, she is my foundation I would not be the person I am without you.

## JORNAL PUBLICATIONS AND POSTER PRESENTATIONS

Parts of this thesis were previously published.

Journal Publications

Aguirre Santos, E.A., Schieber, A., & Weber, F. (2018). Site-Specific Hydrolysis of Chlorogenic Acids by Selected Lactobacillus Species. *Food Research International*, 109, 426-432.

Poster Presentations

Aguirre Santos, E.A., Weber, F., & Schieber, A. (2016). Lactic acid Fermentation of Black Carrot Juice for the Production of Pyranoanthocyanins. *Deutscher Lebensmittelchemikertag*. München.

Aguirre Santos, E.A., Weber, F., & Schieber, A. (2014). Application of Lactic Acid Bacteria for the Production of Precursors used in the Synthesis of Pyranoanthocyanins. *Deutscher Lebensmittelchemikertag*. Gießen.

## SUMMARY

Undoubtedly color is an important factor influencing consumer preferences. The long-discussed safety of synthetic colorants and the demand of customers for more healthy and natural products have resulted in research on natural sources to substitute those artificial colorants. Among natural alternatives, anthocyanins are well-known, nevertheless their use is limited to acidic products due to their poor stability at higher pH and elevated temperatures. However, acylation enhances anthocyanins stability while showing more resistance to pH changes preventing hydration. Black carrot juice is a remarkable coloring food product mainly used since it has been reported to be a rich source of highly stable acylated anthocyanins and phenolic acids. Besides the chemical composition, black carrot juice has shown high antioxidant capacity. Additionally, the use does not require a label declaration as E-number. Pyranoanthocyanins are pigments presented in processed fruit products like juices and wines. The enhanced stability of pyranoanthocyanins at different pH-values renders them as good candidates to be used as natural food colorant. They are reaction products of anthocyanins and different compounds from microbial metabolism. They occur in low concentration making their isolation from natural sources nonpractical. Due to the phenolic composition presented by black carrots, they are a good source of precursors for the synthesis of pyranoanthocyanins. These precursors are anthocyanins and hydroxycinnamic acids. As hydroxycinnamic acids are predominantly bound to quinic acid or sugars, cinnamoyl esterase activity is necessary. Therefore, lactic acid bacteria, able to hydrolyze the ester bond and release the free acids were used. Once released, the hydroxycinnamic acids can be converted into vinylphenols by a second group of LAB displaying hydroxycinnamate decarboxylase enzymes. The formed products react with the anthocyanins to pyranoanthocyanins. The ability to metabolize hydroxycinnamic acids (*p*-coumaric, caffeic acid, and ferulic acid) to vinyl phenols by *L. plantarum*; *L. mali* and *L. sake* was investigated.

Quantification of decarboxylation products and pyranoanthocyanins was conducted by UHPLC-DAD. The results show a strain-dependent variability in the metabolism of hydroxycinnamic acids, with *p*-coumaric acid and caffeic acid as the preferred substrates. The highest yield was observed at pH 4.5. Regarding esterase cinnamoyl activity, *L. fermentum*, *L. helveticus*, and *L. reuteri* strains were able to hydrolyze monochlorogenic and dichlorogenic acids into caffeic acid with considerably different specificities. In this study, a presumed isomerase accompanied by an esterase activity might be responsible for the degradation of 4-CQA by *L. helveticus* and *L. reuteri*. While the synthesis of pyranoanthocyanins in black carrot juice showed that in comparison with naturally occurring concentrations found in conventionally processed fruit products, fermentation with *L. reuteri* revealed high conversion rates. The process reported in this study thus demonstrates a very attractive approach to formulate a coloring foodstuff highly stable to be used for the food industry.

## ZUSAMMENFASSUNG

Die Farbe ist zweifelsfrei ein wichtiger Faktor, der die Konsumentenpräferenz beeinflusst.

Die viel diskutierten Sicherheitsbedenken bezüglich der Nutzung von synthetischen Farbstoffen und die Nachfrage von Konsumenten nach gesünderen und natürlicheren Produkten fördert die Forschung nach alternativen Farbstoffen natürlichen Ursprungs. Anthocyane stellen eine bekannte Alternative zu synthetischen Farbstoffen dar. Ihr Einsatzgebiet ist aufgrund der geringen Prozessstabilität jedoch auf Produkte mit einem hohen Säuregehalt beschränkt. Durch eine Acylierung kann die Stabilität der Anthocyane erhöht und die Empfindlichkeit gegenüber Änderungen im pH-Wert reduziert werden, was eine Hydratation der Anthocyane verhindert.

Schwarzer Karottensaft wird bereits als Färbendes Lebensmittel eingesetzt, da er reich an hochstabilen acylierten Anthocyanen und Phenolsäuren ist. Zudem weist er eine hohe antioxidative Kapazität auf. Für die Anwendung in Lebensmitteln ist er attraktiv, da er auf dem Produkt nicht mit einer E-Nummer deklariert werden muss.

Pyranoanthocyane sind Farbpigmente, die in verarbeiteten Fruchtprodukten vorkommen, wie z.B. in Säften oder in Weinen. Sie eignen sich für die Verwendung als Lebensmittelfarbstoff, da sie eine hohe Stabilität bei verschiedenen pH-Werten aufweisen. Im mikrobiellen Organismus entstehen Pyranoanthocyane als Reaktionsprodukte bei der Reaktion von Anthocyanen mit Metaboliten des mikrobiellen Stoffwechsels. Eine direkte Gewinnung aus natürlichen Quellen ist aufgrund ihrer niedrigen Konzentration möglich. Vorstufen für die Gewinnung von Pyranoanthocyanen sind Anthocyane und u.a. Hydroxyzimtsäuren. Beide finden sich neben anderen phenolischen Inhaltsstoffen in der schwarzen Karotte. Hydroxyzimtsäuren liegen in dieser zum größten Teil an

Chinasäure oder Zucker gebunden vor. In der vorliegenden Studie wurden die Hydroxyzimtsäuren durch die Zimtsäure-Esterase in den Milchsäurebakterien (MSB) *L. fermentum*, *L. helveticus* und *L. reuteri* freigesetzt. Die freien Hydroxyzimtsäuren wurden wiederum durch die Hydroxyzimtsäure-Decarboxylase der MSB *L. plantarum*; *L. mali* und *L. sakei* zu Vinylphenolen decarboxyliert. Reagieren die Vinylphenole mit Anthocyanen, so entstehen Pyranoanthocyane.

Bei der Umsetzung der Hydroxyzimtysäuren zu Vinylphenolen wurden die Decarboxylierungsprodukte und Pyranoanthocyane mittels UHPLC-DAD quantifiziert. Die Ergebnisse zeigen, dass ein Unterschied in der Metabolisierung der Hydroxyzimtsäure durch die verschiedenen MSB-Stämmen besteht. Die bevorzugten Substrate sind *p*-Cumarsäure und Kaffeesäure. Die höchste Ausbeute wurde bei pH 4.5 erzielt. Bezüglich der Zimtsäure-Esterase wurde bei der Hydrolyse von Monochlorogensäure und Dichlorogensäure in Kaffeesäure ein signifikanter Unterschied in der Spezifität zwischen den verschiedenen MSB-Stämmen festgestellt. Es wird angenommen, dass die Degradation von 4-CQA in *L. helveticus* und in *L. reuteri* durch eine Isomerase-Aktivität kombiniert mit einer Esterase-Aktivität verursacht wird. Die Synthese von Pyranoanthocyanen in schwarzem Karottensaft durch *L. reuteri* resultierte in einer deutlich höheren Konzentrationen der Pyranoanthocyane, als man sie in prozessierten Fruchtprodukten findet.

Die vorliegende Studie beschreibt somit einen sehr attraktiven Ansatz für die Gewinnung von stabilen Farbstoffen natürlichen Ursprungs, die in der Lebensmittelindustrie verwendet werden können.

## CONTENTS

## ABBREVIATIONS

| | |
|---|---|
| 3-CQA | 3-Chlorogenic acid |
| 4-CQA | 4-chlorogenic acid |
| 5-CQA | 5-chlorogenic acid |
| 4-VC | 4-Vinylcatechol |
| 4-VG | 4-Vinylguaiacol |
| 4-VP | 4-Vinylphenol |
| 4-EC | 4-Ethylcatechol |
| 4-EG | 4-Ethylguaiacol |
| 4-EP | 4-Ethylphenol |
| ADHD | Attention Deficit Hyperactivity Disorder |
| BCJ | Black carrot juice |
| CA | Caffeic acid |
| CD | Cinnamoyl decarboxylase activity |
| CNA | Cinnamic acid |
| CQA | Chlorogenic acids |
| di-CQAs | dichlorogenic acids |
| EC | Esterase cinnamoyl activity |
| FA | Ferulic acid |
| GRAS | Generally Recognized as Safe |
| GYM | Glucose, Yeast and Malt |
| HCA | Hydroxycinnamic acids |
| L-Tyr | L-Tyrosine |
| L-Phe | L-Phenylalanine |
| MRS | de Man Rugosa and Sharpe |
| PAL | Phenylalanine ammonia lyase |
| *p*-CA | *p*-Coumaric acid |
| PhA | Phloretic acid |
| TAL | Tyrosine ammonia lyase |
| UHPLC-DAD | Ultra-High-Performance Liquid Chromatography |
| UHPLC-ESI-$MS^n$ | Ultra-High-Performance Liquid Chromatography with Diode array and Electrospray Ionization-Mass Spectrometry |
| HPCCC | High performance countercurrent chromatography |
| QA | Quinic acid |

# 1. INTRODUCTION

Color plays an important role in life. It can carry important physiological and psychological links in humans, since color evokes some expectations e.g., triggering perception of taste, stimulating the appetite, sending signals for choosing a product (Clydesdale, 1991; Downham & Collins, 2000; Dufossé, 2006; Lakshmi, 2014). Color has always served as a quality indicator for food used to determine the acceptability of a product. A colorant may be of natural or synthetic origin, they are compounds capable of imparting color when they are added to food. Therefore, the development of attractive products through color is of great importance for the food industry (Dufossé, 2006; Aberoumand, 2011; Hisano, 2016b).

The use of food colorants is far from being a modern development. In fact, this practice goes back to ancient Egypt, where saffron was used to color food (Joshi et al., 2003; Aberoumand, 2011; Esteves Torres et al., 2016). Until the late nineteenth century, the use of pigments derived from animals, colored minerals, spices and plant extracts known as natural colorants, e.g. saffron, indigo, and cochineal, were the main sources for coloring foodstuffs (McKone, 1991; Downham & Collins, 2000; Harasym & Bogacz-Radomska, 2016). In 1856 with the development of Mauveine as the first synthetic colorant by the British chemist William Henry Perkin allowed the world of colorants to spread rapidly (Sousa et al., 2008; Cañamares & Lombardi, 2015; Hisano, 2016b). Following this achievement, some chemistry companies started producing colorants including those for the use in foodstuffs. Nevertheless, some of them were toxic e.g., copper sulphate, indigo or mercury sulphide (McKone, 1991; Downham & Collins, 2000; Hisano, 2016a; Esteves Torres et al., 2016). Regulations regarding the safety of food colorants did not exist during that time. It was in 1906 when the declaration of the use of colorants in food was placed under governmental supervision with the creation of the Pure Food and Drug Act against Food Adulteration in the US (Hisano, 2016a, 2016b). With the creation of this act, an extensive review of food additives began to assess their safety and avoid mislabeled products to ensure the protection of consumers. Moreover, in the 1960s, the E number system was introduced. This system permitted the use of those additives which had been previously permitted by the corresponding authorities. With the introduction of this system, consumers' concern regarding the safety of additives temporarily decreased (Saltmarsh, 2014). In the 1980s, a full declaration of ingredients and additives used in a food product became obligatory. Due to dissent from countries, no uniform regulation for colorants exists. Meanwhile, clean labeling has gained increasing interest and has become a trend in the food industry. This term defines labels where all ingredients are not from an artificial origin (Downham & Collins, 2000; Saltmarsh, 2014). Thus, interest in clean labeling is continuously growing in the food market, since consumers try to avoid products containing synthetic additives, particularly colorants and preservatives (Downham & Collins, 2000).

Today, within the European Union (EU), it is obligatory that all additives used in a product must be declared. The EU has authorized the use of 43 colorants (European Directive 94/36/EC) and this authorization ensures that adverse health effects and quantities of use are strictly regulated. Each colorant has been assigned an E number, whereby 17 are synthetic colorants and 26 are natural or naturally derived (Downham & Collins, 2000; Mortensen, 2006; Aberoumand, 2011). All of them are regulated without making any distinction among their source of origin, which is, synthetic (e.g. azo-dyes) or natural (e.g. anthocyanins).

In the food industry, colorants are added to processed food e.g., soft drinks, candies, margarine, cheese, jam, and confectionery. In 2016, the total colorant market size was estimated at $1.3 billion, while the demand for natural colorants in 2020 is expected to reach $1.7 billion. North-America and Europe are the largest markets for natural colorants (Harasym & Bogacz-Radomska, 2016).

Among the natural colorants, anthocyanins, betalains, caramel, carminic acid, carotenoids, chlorophylls, curcuminoids, and minerals can be named. Anthocyanins are Generally Recognized as Safe (GRAS) (Aberoumand, 2011; Harasym & Bogacz-Radomska, 2016). The interest in anthocyanins has been increasing since they have been reported to possess positive health effects. Antioxidant, anti-inflammatory, and anti-degenerative activities are some of the biological properties that have been associated with them. In the food industry, anthocyanins have been used as food colorants for a long time in many products, e.g., jams and beverages (Downham & Collins, 2000; Shipp & Abdel-Aal, 2010). Nevertheless, techno-economic factors limit their use, such as their low stability and the higher productions costs compared to artificial colorants.

The food industry is forced to find a way to substitute artificial colorants with natural counterparts (Aberoumand, 2011). The application of biotechnology brings an opportunity for the synthesis of natural colorants. The use of microorganisms in the food industry plays an important role due to their high potential in multiple applications, easy production and easy downstream processing (Wissgott & Bortlik, 1996; Dufossé, 2006; Mapari et al., 2010; Venil et al., 2013; Dufossé et al., 2014).

In 1996, Cameira-dos-Santos et al. found two novel anthocyanin derivatives. These compounds were products from the reaction between the major anthocyanins in wine, i.e., malvidin 3-monoglucoside and malvidin 3-(6-*p*-coumaroyl)monoglucoside with 4-vinylphenol (4-VP). In comparison with anthocyanins, these new pigments showed higher stability and a wider range of colors even at higher pH. These pigments are present in processed food and they occur as a product of the direct reaction between anthocyanins and small molecules such as pyruvic acid and hydroxycinnamic acids resulting in a new pyran ring. The properties displayed by these pigments known as pyranoanthocyanins make them promising candidates to be used as food colorants. However, the recovery of pyranoanthocyanins from natural sources is limited due to the low quantities in which they can be found (Vallverdú-Queralt et al., 2016). Pyranoanthocyanins are reaction products of anthocyanins and different compounds from microbial metabolism (Cameira-dos-Santos et al., 1996; Fulcrand et al., 1996; Hayasaka & Asenstorfer, 2002; Wang et al., 2003). Colorants required a previous extraction to be used. In contrast, a coloring foodstuff can be defined as a food extract or concentrate able to be used as food with coloring effects. Within the EU, a coloring foodstuff may be labeled as an ingredient appearing mainly as concentrates. Therefore, an E label is not required. The substitution of colorants can be accomplished by the use of coloring foodstuff, which, for the food industry, can be useful in trying to avoid the use of E additives and offering an eco-friendly alternative for the consumers. Among sources of natural colorants, black carrot juice (BCJ) is known for its application as coloring foodstuff (Downham & Collins, 2000; Wrolstad & Culver, 2012; Lakshmi, 2014). The phenolic composition in black carrots offer a good source of precursors for the synthesis of pyranoanthocyanins. In black carrots, hydroxycinnamic acids can be found as natural compounds (Kammerer et al., 2003). They can play an important role as copigments to improve anthocyanins' stability. Furthermore, they can act as precursors for the formation of more stable pyranoanthocyanins. So far, the development of protocols for the synthesis of pyranoanthocyanins in BCJ and their further use as food coloring is lacking.

# 2. BACKGROUND

## 2.1. Food colorants

The color presented by natural or processed products is an attribute that influences their consumer acceptance. Color is described as a property that consumers associate with taste, nutritional value and the overall quality of a product.

The use of colorants can be traced back to ancient times. There are reports of the Egyptians (400 BC) using natural pigments for food, drugs, and cosmetic purposes. The introduction of synthetic colorants was made in 1856 with "Mauve", by the British chemist William Henry Perkin while experimenting to find a cure for malaria. Since the discovery of this colorant, many other synthetic dyes were used to substitute natural dyes in an uncontrolled way (Sousa et al., 2008).

Pigments or dyes are terms often used to describe a colored substance. The difference between both is their size and solubility. Pigments are insoluble in the medium, whereas a dye is soluble (Sousa et al., 2008; Cañamares & Lombardi, 2015; Esteves Torres et al., 2016).

Food colorants are mainly used to mitigate color losses occurring during processing, to bring color to uncolored products, to improve the natural color, or to standardize a product from batch-to-batch thus improving their acceptability and making them more appetizing (Aberoumand, 2011). The colorants used in foodstuffs may be classified into four main categories: (1) natural colorants, (2) natural identical colorants, (3) synthetic colorants, and (4) inorganic colorants (Mortensen, 2006; Mapari et al., 2010; Aberoumand, 2011).

As mentioned by Dabas et al. (2011), natural colorants can be defined as natural pigments or pigments made by living organisms. Sources of them can be plants, animals, or different microorganisms. Phenolic compounds like anthocyanins, as well as numerous other compounds are naturally occurring colorants. Among these natural colorants, carotenoids and anthocyanins are probably the most known and investigated (Mazza & Brouillard, 1987; Malien-Aubert et al., 2001; Schwarz & Winterhalter, 2003; Wu & Prior, 2005; Esteves Torres et al., 2016). Rich sources of anthocyanins like blackberries, grapes, raspberries, elderberries, red cabbage, and black carrots are already used by the food industry for extracting and isolating these natural pigments (Mortensen, 2006). However, the use of anthocyanins in food is limited due to their low stability affected by several factors, and difficulties in their extraction and purification (Baublis et al., 1994; Sarni-Manchado et al., 1996; Giusti & Wrolstad, 2003; Schwarz & Winterhalter, 2003; Ignat et al., 2011; Sigurdson et al., 2017). Colors resulting from modification on the material by living organisms may also be included in this category, i.e., biocolors that cannot be found in nature. Pyranoanthocyanins are a good example of biocolors, their formation occurs during wine ageing, were microbial products react with anthocyanins acting as precursors for these new pigments. The production of carotenoids, flavonoids and other compounds by microorganisms is well documented in literature (Dufossé, 2006; Chattopadhyay et al., 2008; Wrolstad & Culver, 2012; Heer & Sharma, 2017). As mentioned by Heer and Sharma (2017), the use of pigments produced by microorganism is a recent trend. The biosynthesis of these compounds can be easily scaled up since, compared to the cultivation of edible plants, the process is not weather dependent, and the production conditions can be standardized. Therefore, biotechnological production of such colorants represents an attractive alternative. Cochineal is maybe the most known insect origin pigment belonging to this group. With its origin in south and central America, the red pigment produced by insects has been applied for dying of textiles, food products and cosmetics (Madsen et al., 1993; Méndez-Gallegos et al., 2003; Takaichi et al., 2003; Borges et al., 2012; Velmurugan et al., 2013). Examples of microorganisms that are known to produce biocolors are *Staphylococcus aureus, Serratia marcescens* (Heer & Sharma, 2017), and

*Vibrio gazogenes* (Alihosseini et al., 2008), the algae *Dunaliella salina* (Pisal & Lele, 2005; Spolaore et al., 2006), the yeast *Rhodotorula* sp. *Rhodotorula glutinis* (Hernández-Almanza et al., 2014), and *Phaffia rohodozyma* (Johnson & Lewis, 1979; An et al., 1989), and the fungi *Monascus purpureus* (Fabre et al., 1993; Akihisa et al., 2005; Lakshmi, 2014).

Regarding natural identical colorants, these compounds are produced by chemical synthesis, replicating the molecular structure of naturally occurring compounds. Synthetic colors are compounds like azo-dyes that are not found in nature. After the discovery of mauvine in 1856, many others were synthetized revolutionizing the colorant industry and opening up new opportunities for processed food. In the late nineteenth century, natural colorants were almost completely replaced by synthetic colorants.

There are some factors affecting the stability of coloring agents, e.g., temperature, light, or other components found in the media like oxidizing and reducing agents (Joshi et al., 2003; Mortensen, 2006; Socaciu, 2008). Since their discovery, synthetic colorants have been used more extensively due to their properties, e.g., higher stability, color intensity, hue, and most probably their low production costs. However, there are just a few natural colorants that can be found in quantities abundant enough to make their application feasible. In addition, natural colorants have been shown to be less efficient compared to the artificial ones, suggesting that higher quantities of the natural colorants need to be added to reach the desired color. This, in turn, can directly affect the flavor, stability, and the cost of the product (Mortensen, 2006; Bechtold & Mussak, 2009; Wrolstad & Culver, 2012).

In recent years, consumer awareness has increased, as several studies have reported side effects of synthetic colorants e.g., toxicological effects, carcinogenic effects, allergic reactions, and the well-investigated attention deficit-hyperactivity disorder (ADHD) (Combes & Haveland-Smith, 1982; Sasaki et al., 2002; Joshi et al., 2003; Schab & Trinh, 2004; Bateman et al., 2004; Wang et al., 2006; McCann et al., 2007; Pan et al., 2009; Sharma et al., 2011; Elbanna et al., 2017). Probably, the most extensive study regarding artificial colorants was published by the University of Southampton. The study reported a detrimental effect of six artificial food colorants children's behavior. While the UK prohibited the application of these six colorants in food, the EU only requests that a label warning of potential adverse effects in children be added.

Therefore, the challenge for the food industry is the substitution of synthetic dyes with colorants obtained from natural sources. Moreover, and as suggested by the regulation of the European Commission 13333/2008 to avoid the use of food additives, the coloring food market will probably change toward more natural options like coloring foodstuffs. Actually, extracts from aronia, black carrot, blackcurrant, and hibiscus can be found in the food industrial market. Table 2.1 shows an example given by SMARTIES® which successfully avoids using artificial colorants through implementing coloring foodstuffs and clean label (Bobe & Michel, 2011).

**Table 2.1**. Colorants used for SMARTIES®

| Colorants used in 2006 | Colorants used in 2009 |
|---|---|
| E104 Quinoline yellow | Lemon (yellow) |
| E110 sunset yellow | Safflower (yellow) |
| E122 Carmoisine | Radish (red) |
| E120 Carmine | Red cabbage (red) |
| E133 Brilliant blue | Hibiscus (red) |
| E124 Ponceau 4R | Spirulina (blue) |
| | Black carrot (purple) |

Thus, the implementation of coloring foodstuffs can be simplified because no extraction techniques are required. In addition, using coloring food stuffs adds value to the food, thanks to their potential health benefits and their application can be standardized over the world.

## 2.2. Anthocyanins

The word anthocyanin derives from two Greek words: *anthos*, flower and *kyanos*, blue. Anthocyanins belong to the family of flavonoids and are water-soluble pigments causing the red, pink, blue, or purple colors of many plants, i.e., flowers, fruits, and vegetables (Wu & Prior, 2005; Castañeda-Ovando et al., 2009; Motilva et al., 2013; Quina & Bastos, 2018). These compounds are located in vacuoles of plant cells acting as a defense against pathogens and predators, attracting animals for pollination and seed dispersal, providing plant pigmentation, acting as light screen against harmful radiation and as antioxidants (Stintzing & Carle, 2004). Due to the range of attractive colors varying from pink to purple hues, they are used as natural food colorants (Bakker & Timberlake, 1997; Stintzing & Carle, 2004; Bordignon-Luiz et al., 2007; Khoo et al., 2017; Quina & Bastos, 2018). Anthocyanins represent an alternative in the food industry to substitute synthetic colorants. Many edible plants are source of anthocyanins including, purple sweet potato, red grapes, black carrot, blackcurrant, and elderberry (Downham & Collins, 2000; Wrolstad & Culver, 2012; Lakshmi, 2014).

Anthocyanin extracts are used in the food industry as additives in beverages, jellies, confectioneries and jams. All anthocyanins are used under classification E 163 number of the EU legislation of food additives (Wrolstad & Culver, 2012). The application of anthocyanins in the EU is at *quantum satis* level, i.e., no maximum level is specified, with exception of fruit-flavored breakfast cereal where a maximum of 200 ppm is established (Hendry & Houghton, 1996; Frick, 2003; Bechtold & Mussak, 2009). Nevertheless, their application in the food industry is limited due to their poor stability shown at low acid conditions, occurring during processing and storage. However, anthocyanins are of scientific interest due to their potential health benefits, which include acting as antioxidant, anti-inflammatory, anti-tumor, neuroprotective, and chemopreventive agents

(Prior, 2003; Kong et al., 2003; Tarozzi et al., 2007; Yao et al., 2010; Bishayee et al., 2011; 2011b; Sun et al., 2012; Khoo et al., 2017).

The basic structure of anthocyanins are the anthocyanidins (2-phenylbenzopyrylium or flavylium salts), which consists of a C6-C3-C6 skeleton with an aromatic ring A bonded to a heterocyclic ring C containing oxygen and linked by a carbon-carbon bound to a second aromatic ring B (Figure 2.1). The principal differences between the anthocyanidins are the hydroxy or methoxy substituents at the A and B rings. More than 90% of all anthocyanins identified are based on six anthocyanidins, i.e., pelargonidin, cyanidin, peonidin, delphinidin, petunidin, and malvidin (Mazza & Miniati, 1993; Kong et al., 2003; Mercadante & Bobbio, 2008; Fernández-López et al., 2013; Quina & Bastos, 2018).

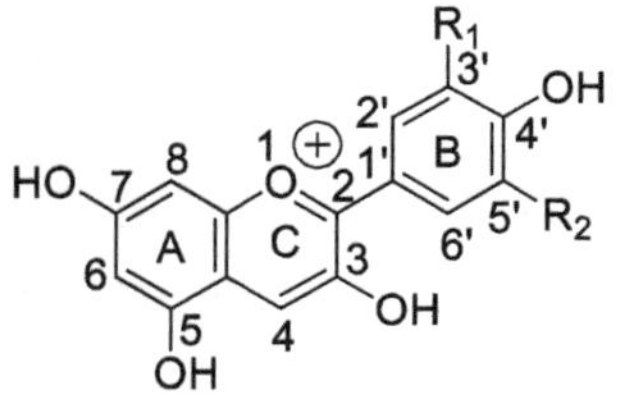

| Anthocyanidin | $R_1$ | $R_2$ | Λ max [a] |
|---|---|---|---|
| Pelargonidin | H | H | 494 |
| Cyanidin | OH | H | 506 |
| Delphinidin | OH | OH | 508 |
| Peonidin | $OCH_3$ | H | 506 |
| Petunidin | $OCH_3$ | OH | 508 |
| Malvidin | $OCH_3$ | $OCH_3$ | 510 |

[a] λ max given in nm

**Figure 2.1.** Structures of the most abundant anthocyanidins

In nature, anthocyanidins mainly occur as glycosides. The sugars that are commonly bound to the anthocyanidins are glucose, galactose, arabinose, rhamnose, and xylose. Glycosylation renders more stable anthocyanins compared to their respective aglycones. They can primarily occur at positions 3, 5 and 7 (Wrolstad, 2004), moreover, it is also possible to find glycosylations at position 3', 5' and 4' (Brouillard, 1988; Mazza & Miniati, 1993). In addition, anthocyanins can be acylated with aliphatic or aromatic acids attached to the sugar moieties (Figure 2.2). Among the acyl derivatives, hydroxycinnamic acids, i.e., caffeic (CA), ferulic (FA), *p*-coumaric (*p*-CA), and sinapic acids, can be found as well as aliphatic acids like acetic, malic, malonic, oxalic, and succinic acids (Mazza & Brouillard, 1987; Mazza & Miniati, 1993; Clifford, 2000a). It is also possible to find anthocyanins presenting multiple acyl groups attached to the molecule.

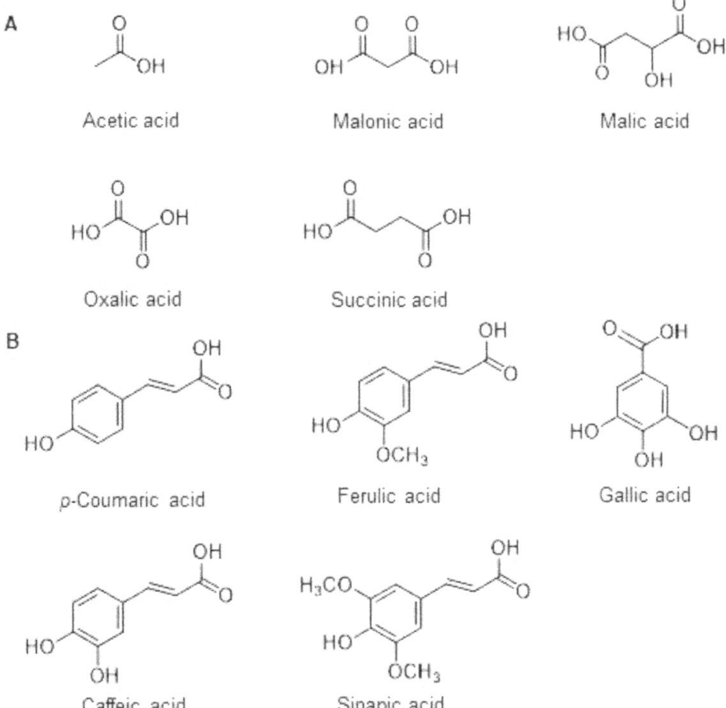


**Figure 2.2.** Common aliphatic acids (A) and hydroxycinnamic acids (B) attached to sugar moieties of anthocyanins

The kind and number of sugars as well as acylations and their respective position are responsible for anthocyanins' stability and reactivity (Mazza & Miniati, 1993). In addition, the number of hydroxy and methoxy groups of the aglycone influences the color presented, whereby an increment in hydroxylation results in a blue shift, while an increment in the number of methoxy groups leads to red colors (Mazza & Brouillard, 1987; Mateus & Freitas, 2001; Mercadante & Bobbio, 2008). However, the most important factor determining anthocyanins' color is defined by the pH-value of the media, since anthocyanins can exist in different colored and colorless forms. Due to all these possible variations of their substitution pattern, more than 600 anthocyanins have been reported so far in the plant kingdom (Clifford, 2000a; Wu & Prior, 2005).

### 2.2.1. Factors affecting the stability of anthocyanins

In addition to the inherent effect of the anthocyanin structure on their stability, there are factors like enzymes, light, oxygen, temperature, copigments, or metal ions which have a big influence on their stability (Brouillard, 1983; Mazza & Brouillard, 1990; Mazza & Miniati, 1993; Rodriguez-Saona et al., 1999; Malien-Aubert et al., 2001). The extent of influence of these factors, in turn, depends on the source and the structures of anthocyanins.

#### 2.2.1.1. pH

The pH of the media is directly responsible for the change of anthocyanins' coloration, since several reactions can take place such as proton transfer, hydration, isomerization, and tautomerization (Brouillard, 1988; Basílio & Pina, 2016; Quina & Bastos, 2018). The study of the structural pH-dependent transformation of anthocyanins is of great importance in order to use them as food colorants. Anthocyanins present their greatest stability at acidic pH. Figure 2.3 shows the structural changes that anthocyanins undergo in aqueous solution at different pH-values. Between pH 1 and 2, the red flavylium cation is the predominant species and anthocyanins are highly stable. The flavylium cation undergoes hydration when the pH is raised to 3 and the colorless carbinol pseudobase (A) and yellowish chalcone (B) are formed by a nucleophilic attack of water at C-2. Increasing the pH to values up to 6 leads to deprotonation yielding to the purple quinoidal bases (C) (Timberlake & Henry, 1986; Mazza & Brouillard, 1987; Sarni-Manchado et al., 1996; Horbowicz et al., 2008; Sigurdson et al., 2017). Thus, based on these transformations of anthocyanins at different pH values, anthocyanins' applications are limited to acidic food to ensure the flavylium cation as predominant species and avoid discoloration products occurring at pH higher than 4.

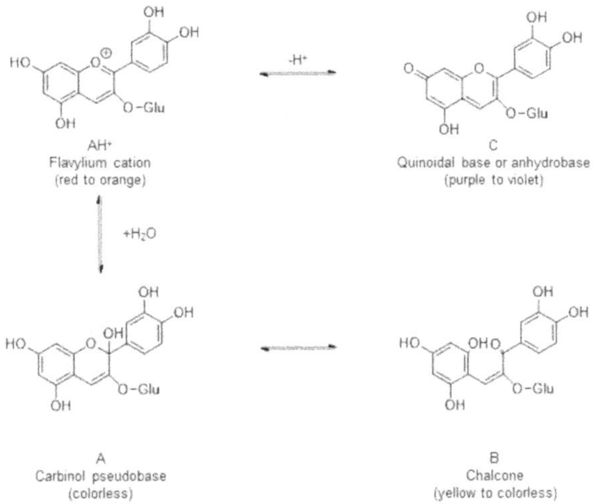

$AH^+$
Flavylium cation
(red to orange)

$-H^+$

C
Quinoidal base or anhydrobase
(purple to violet)

$+H_2O$

A
Carbinol pseudobase
(colorless)

B
Chalcone
(yellow to colorless)

**Figure 2.3.** Effect of pH-value on Cy-3-Glu structure

#### 2.2.1.2. Copigmentation

In nature, the coloration and stability exhibited by anthocyanins cannot be explained by the structural transformations at different pH ranges alone. Anthocyanins also can improve their stability through associations with other compounds in a process called copigmentation.

As a result of these copigmentation associations, higher color intensity is exhibited by anthocyanins than by their free forms (Eiro &

Heinonen, 2002; Trouillas et al., 2016; Khoo et al., 2017). Nevertheless, such associations cannot be predicted, since they are conditioned by other factors, e.g., the concentration in which anthocyanins are found, the ratio between the anthocyanins and the copigment, the nature of the copigment and the pH-value (Mazza & Brouillard, 1987; Brouillard & Dangles, 1994; Boulton, 2001). Among the latter, this process has been reported as a pH-dependent process taking place favorably at pH-values where the flavylium cation can be found as the predominant species (Brouillard & Dangles, 1994; Trouillas et al., 2016).

These complexes occur by intermolecular interaction with other compounds or by self-association with other anthocyanins. There is an extensive list of compounds such as flavonoids, alkaloids, organic acids, and amino acids that can act as copigments (Mazza & Brouillard, 1987; Boulton, 2001; Eiro & Heinonen, 2002; Escribano-Bailon & Santos-Buelga, 2012; Quina & Bastos, 2018). Hydroxycinnamic acids, i.e., CA, FA, *p*-CA, chlorogenic acid (CQA), and sinapic acid are reported to act efficiently as copigments (Bakowska-Barczak, 2005; Castañeda-Ovando et al., 2009). As reported by Wrolstad et al. (2005), the increase of glycosidic substitutions and sugar's acylations with cinnamic acids, improve the anthocyanin stability. A cofactor is regarded as being efficient if it provokes a hyperchromic and bathochromic shift resulting in a color enhancement. Besides that, anthocyanins can act as copigments by increasing their stability through self-association (Asen et al., 1972). Anthocyanins as copigment for other anthocyanins, and, intermolecular copigmentation have been extensively studied by authors like Eiro & Heinonen, 2002; Bordignon-Luiz et al., 2007; Del Pozo-Insfran et al., 2007; Mollov et al., 2007, and, Goto & Kondo, 1991; González-Manzano et al., 2008, respectively. It is proposed that self-association of anthocyanins happens through hydrophobic interaction between the anthocyanin nuclei stacked parallel to each other (González-Manzano et al., 2008; Trouillas et al., 2016). The self-association occurs in highly concentrated solutions, i.e., > 1 mM (Boulton, 2001; González-Manzano et al., 2008). In addition, this reaction is characterized by a hypsochromic shift in the wavelength of the absorbance maximum towards lower values (Mazza & Brouillard, 1990; González-Manzano et al., 2008; Trouillas et al., 2016).

Copigmentation reactions are produced by intramolecular associations when the pigment and copigment are part of the same molecule. Here the associations are between the pyrylium ring of the flavylium cation with their organic acids, aromatic acyl groups, or a flavonoid or a combination of them (Goto & Kondo, 1991; Figueiredo et al., 1996). Intermolecular and intramolecular associations are presented as sandwich-type stacking through hydrophobic and π-π interactions also known as charge-transfer complexes between molecules with rich π-system and the anthocyanin. These complexes improve anthocyanins' stability while hindering the nucleophilic attack of water which leads to colorless forms (Brouillard & Dangles, 1994; Figueiredo et al., 1996; Quina & Bastos, 2018). As reported by Dangles et al., (1993), intramolecular copigmentation is more efficient than intermolecular. As a result of the formation of these complexes, a hyperchromic effect with an increase in the absorption maxima is produced and causes a bathochromic shift (Mazza & Brouillard, 1987; Boulton, 2001; Wu & Prior, 2005; Bordignon-Luiz et al., 2007; Castañeda-Ovando et al., 2009; Solymosi et al., 2015). In figure 2.4, the different mechanisms of copigmentation are shown.

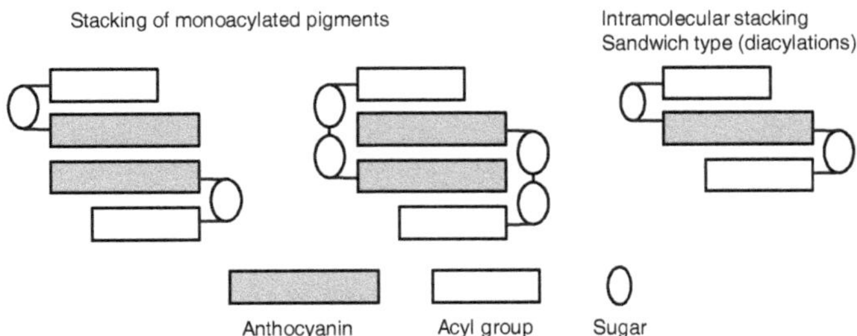


**Figure 2.4.** Stabilization of anthocyanins by copigmentation associations (Giusti and Wrolstad, 2003)

Anthocyanin applications have been achieved with the use of copigments, resulting in an increase of the half-life of a product and decreased degradation of anthocyanins during processing (Rein et al., 2005; Bordignon-Luiz et al., 2007; Del Pozo-Insfran et al., 2007; Mollov et al., 2007). Rich sources of acylated anthocyanins have been shown to be more stable compared to their non-acylated analogs, making them promising candidates for use as food colorants (Giusti & Wrolstad, 2003; Bakowska-Barczak, 2005; Khoo et al., 2017). As reported by Hayashi et al., (1996), after isolating anthocyanins from different edible plants, anthocyanins' composition affects anthocyanins' stability. These authors reported that acylated anthocyanins displayed higher stability under UV irradiation derived from intramolecular associations. This intramolecular association presented by acylated anthocyanins can explain the unusual stability at neutral or weakly acidic conditions.

#### 2.2.1.3. Temperature

Temperature is an important factor influencing the stability of anthocyanins (Zhong & Yoshida, 1993; Zhang et al., 1997; Mori et al., 2005). In the food industry, thermal processing means the use of temperatures ranging from 50 °C to 150 °C, where losses during processing and storage are inevitable (Patras et al., 2010). Anthocyanins are labile pigments, unstable at higher temperatures (Delgado-Vargas et al., 2000; Morais et al., 2002). In addition, temperature combined with increases in the pH can result in the hydrolysis of glycosidic bond, which increases the degradation rate of anthocyanins, due to the reduced stability of the aglycones compared to their glycosylated forms. Degradation of anthocyanins has been reported to follow first order reaction kinetics where chalcones have been proposed as a first product (Adams, 1973; Delgado-Vargas & Paredes-Lopez, 2002; Sadilova et al., 2006; Kırca et al., 2006; Horbowicz et al., 2008). Sadilova et al. (2006) conducted experiments regarding the effect of temperature on the degradation of anthocyanins. These authors exposed juices from strawberries, elderberries, and black carrot to temperatures of 95 °C for 1 to 7 h. At these temperatures, a significant anthocyanin decrease was observed during the first 3 h for all the juices. After 3 h heating, 50 % of elderberry anthocyanins were retained. Furthermore, these authors mentioned a greater anthocyanin stability with respect to black carrot anthocyanins, which is mainly based on the acylation. These results are in agreement with those found by Kirca et al. (2006), who reported a high stability of black carrot anthocyanins under temperatures ranging from 70 °C to 90 °C. Storage temperature showed a fast decrease of anthocyanins at 37 °C, while at refrigerated conditions (4 °C), an increase in the stability was observed. The effect of storage temperature on anthocyanins' stability was also studied by García-Viguera et al. (1999). These authors reported significant losses of anthocyanins in jam during storage at 20 °C, 30 °C, and 37 °C. Higher degradation rates were observed at higher temperatures. This study also confirmed that formation of polymeric pigments increases during storage and heat treatments. This is in agreement with several studies (Ochoa et al., 1999; Clifford, 2000a; Ahmed et al., 2004; Wrolstad et al., 2005; Cisse et al., 2009). Higher stability of anthocyanins can be accomplished during processing and storage by using short treatment times at low temperatures (Patras et al., 2010).

#### 2.2.1.4. Oxygen, enzymes, and light

Although the effect of pH is more important for the stability of anthocyanins, oxygen is one factor that accelerates anthocyanin degradation depending of the pH value (Attoe & Elbe, 1981; Delgado-Vargas et al., 2000; Bordignon-Luiz et al., 2007). The presence of oxygen can be detrimental to anthocyanins causing their oxidation and yielding brown or colorless compounds. Enzymes present in plant tissues, i.e., glycosidases, polyphenoloxidases and/-or peroxidases, make anthocyanins susceptible to degradation. In presence of oxygen, polyphenol oxidase is the dominant enzyme influencing anthocyanin degradation (Delgado-Vargas & Paredes-Lopez, 2002; Jaiswal et al., 2010; Patras et al., 2010).

Moreover, light combined with high temperatures may also accelerate their degradation (Attoe & Elbe, 1981; Sarni-Manchado et al., 1996; Bordignon-Luiz et al., 2007). Degradation of anthocyanins by exposure to light depends on the oxygen present (Attoe & Elbe, 1981). Authors like Attoe & Von Elbe (1981), evaluated the effect of fluorescent light in presence of oxygen at different ranges of temperatures. These authors reported that at higher temperatures, degradation of anthocyanins was mainly attributed to thermal destruction. Moreover, at temperatures higher than 40 °C, light played an important role in the degradation of anthocyanins. Fluorescent light had no effect in the absence of oxygen. Similar studies were carried out by Bordignon-Luiz (2007) in the analysis of color stability of anthocyanins from Isabel grapes. In this study, the impact of temperature, pH, presence or absence of light, and oxygen was evaluated. Samples kept in darkness exhibited a lower decrease in anthocyanin content than those in the presence of light. Moreover, the degradation of anthocyanins was significant in the presence of oxygen. Zheng et al. (2007) studied anthocyanin changes in strawberries stored under air and high oxygen conditions over 14 days. These authors reported significant losses in the strawberries' quality due to their exposure to high oxygen atmospheres. A study evaluating the effect of polyphenol oxidase from dried arils of pomegranate conducted by Jaiswal et al. (2010) reported a loss of 65 % of anthocyanin content in samples exposed to air, while pure anthocyanins proved to be stable at high temperatures in the absence of oxygen. Thus, inactivation of polyphenol oxidase enzymes is possible at high temperatures with short treatment-times avoiding degradation of anthocyanins.

### 2.3. Origin of pyranoanthocyanins

At wine pH, i.e., 3 to 4, anthocyanins are in a highly reactive state (Nagel & Wulf, 1979; Asenstorfer et al., 2001; Monagas et al., 2007; Oliveira et al., 2010). As reported by several authors, a high percentage of free anthocyanins will condense with other phenolic compounds to form more stable pigments, the rest will undergo reactions such as oxidation, degradation, or transformation into colorless compounds (Bakker & Timberlake, 1997; Fulcrand et al., 1996; Vivar-Quintana et al., 1999; Mateus et al., 2003; Alcalde-Eon et al., 2004; He et al., 2006; Monagas et al., 2007; Rentzsch et al., 2009; Oliveira et al., 2010). Anthocyanins can react and form more stable structures through condensation reactions with different compounds, e.g., flavan-3-ols, acetaldehyde, pyruvic acid, and those products from microbial metabolism (Wang et al., 2003; He et al., 2006; Monagas et al., 2007; Vallverdú-Queralt et al., 2016). Pyranoanthocyanins are pigments that were first found by Cameira dos-Santos et al. (1996) in red wine filtrates. They were first isolated during cross-flow microfiltration and were found to be formed during fermentation and ageing of red wines. These pigments were malvidin-derivatives formed by the reaction between the two major wine anthocyanins and 4-VP (Cameira-dos-Santos et al., 1996; Sarni-Manchado et al., 1996).

Pyranoanthocyanins are named after the additional ring D formed at C-4 and C-5 of the native anthocyanin. The first group of these pigments, known as hydroxyphenyl-pyranoanthocyanins, are formed by the reaction of hydroxycinnamic acids or 4-vinylphenol derivatives with anthocyanins. Figure 2.5 shows the formation of these pigments, starting by the cycloaddition of the different precursors with a further dehydration and oxidation (Fulcrand et al., 1996; Rentzsch et al., 2007; de Freitas & Mateus, 2011).

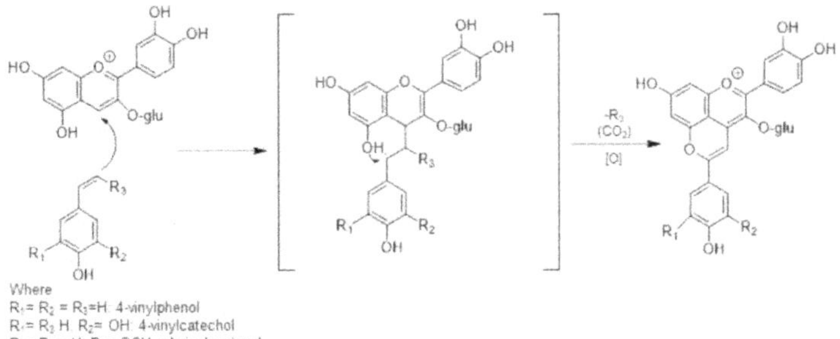


**Figure 2.5.** Reaction pathway of the formation of hydroxyphenyl-pyranoanthocyanins

The influence of anthocyanins and pyranoanthocyanins on the color of aged wines has extensively been studied. During wine ageing, anthocyanins concentration decreases constantly, whereas the color presented is maintained. As a result of this wide variety of reactions, color stability might be mainly attributed to the presence of pyranoanthocyanins (Fulcrand et al., 1998; Mateus & de Freitas, 2001; Wang et al., 2003; He et al., 2006; Monagas et al., 2007). Consequently, the concentration of pyranoanthocyanins can contribute to the shift from red-purple to more red-orange colorations (Fulcrand et al., 1996; Mateus & Freitas, 2001; Alcalde-Eon et al., 2004; He et al., 2006; Rentzsch et al., 2009).

### 2.3.1. Occurrence of pyranoanthocyanins

Pyranoanthocyanins are not present in fresh food material. Studies have shown that they are formed during anthocyanin extraction or during storage of processed food (Fan et al., 2008; Marquez et al., 2013). Since their discovery, numerous pyranoanthocyanins have been characterized. Figure 2.6 shows the most abundant pyranoanthocyanins.

A classification is complicated due to the different compounds that may act as precursors; among them, pyruvic acid, α-ketoglutaric acid, and acetaldehyde (Fulcrand et al., 1996; Bakker & Timberlake, 1997; Vivar-Quintana et al., 1999; Mateus & de Freitas, 2001; Schwarz et al., 2003c; Morata et al., 2003). Other pyranoanthocyanins are products of the reaction between anthocyanins and acetone or acetoacetic acid known as methylpyranoanthocyanins (Lu & Foo, 2001; Hayasaka & Asenstorfer, 2002). Furthermore, it is possible to obtain pyranoanthocyanins from the reaction of anthocyanins and phenolic acids, i.e., *p*-CA, FA, CA, sinapic acid, and their respective vinyl derivatives (Cameira-dos-Santos et al., 1996; Fulcrand et al., 1998; Hayasaka & Asenstorfer, 2002; Schwarz & Winterhalter, 2003; Wang et al., 2003; deFreitas & Mateus, 2011; Gómez Gallego et al., 2013).

Pyranoanthocyanins have been identified and isolated from black currant juice (Lu et al., 2000), blood orange juice (Hillebrand et al., 2004), black carrot juice (Schwarz et al., 2004) and modified berry juices (Rein et al., 2005). Adducts of vinylphenol have been found in sorghum sourdough after fermentation (Bai et al., 2014), nevertheless, most studies have been done with wine or wine models (Fulcrand et al., 1998; Mateus & de Freitas, 2001; Schwarz et al., 2003a; Monagas et al., 2007; Rentzsch et al., 2009; Oliveira et al., 2010), or fortified wines (Bakker & Timberlake, 1997).

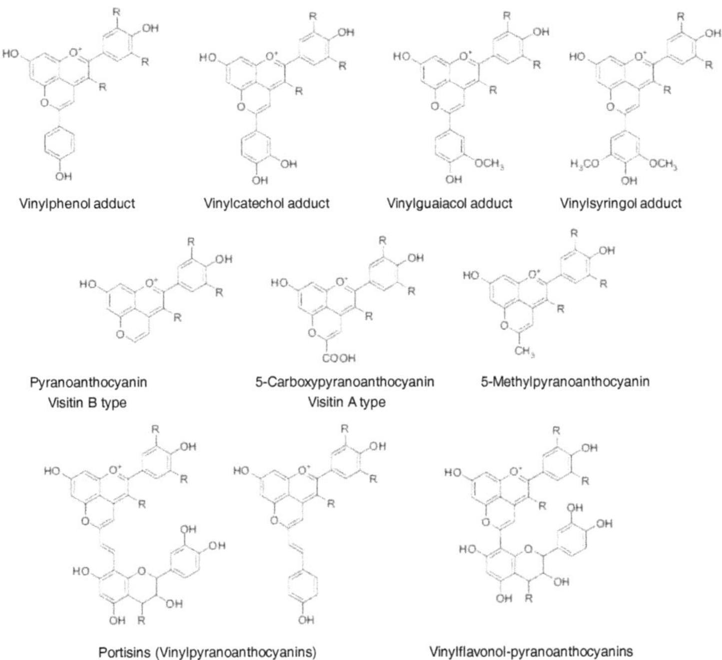


**Figure 2.6.** Structures of the main pyranoanthocyanins (Rentzsch et al., 2007)

Further reactions lead to the modification of pyranoanthocyanins, known as pyranoanthocyanins of second generation, e.g., portisins, oxovisitins, and pyranoanthocyanin dimers (Mateus et al., 2003; Mateus et al., 2004b; He et al., 2010b; Marquez et al., 2013). This second generation of pyranoanthocyanins is a group in which anthocyanins do not directly act as precursors.

#### 2.3.1.1. Origin of hydroxyphenyl-pyranoanthocyanins

In wine, pyranoanthocyanins precursors are found as natural constituents of grapes (Ribéreau-Gayon et al., 2006). Anthocyanins are mainly responsible for the wine color (Mateus et al., 2004a; Marquez et al., 2013). During fermentation, anthocyanin concentration decreases and other more stable compounds can be formed e.g., pyranoanthocyanins (Fulcrand et al., 1996; Schwarz et al., 2003b; Mateus et al., 2004a). These compounds have been shown to contribute to the color change from red-purple to orange hues occurring in aged wine (Mateus & de Freitas, 2001; Vivar-Quintana et al., 2002; Hayasaka & Asenstorfer, 2002).

As described by Gómez Gallego et al. (2013), formation of hydroxyphenyl-pyranoanthocyanins is related to the increase in hydroxycinnamic acids during wine aging. In wine, the presence of vinyl phenols is associated with the formation of off flavors (Chatonnet et al., 1993; Fulcrand et al., 1996; Mateus et al., 2003). Nevertheless, the formation of pyranoanthocyanins causes a beneficial decrease in the concentration of vinylphenols in wine since they act as precursors of pyranoanthocyanins (Boulton, 2001; Monagas et al., 2007; Benito et al., 2009). Vinylphenol derivatives are decarboxylation products of hydroxycinnamic acids formed by cinnamate decarboxylase (CD) activity of some microorganisms, e.g., *Saccharomyces cerevisiae* or *Pichia guillermondii* (Chatonnet et al., 1993; Asenstorfer et al., 2001; Morata et al., 2003; Schwarz & Winterhalter, 2003; Benito et al., 2009; Morata et al., 2012; Vallverdú-Queralt et al., 2016).

Hydroxyphenyl-pyranoanthocyanins have been widely studied in matured wines. Their formation was first reported by Cameira dos Santos et al. (1996); later Fulcrand et al. (1996) contributed by identifying the structure of those new pigments. Hayasaka and Asenstorfer, (2002) screened wine extracts, and identified, among others, anthocyanins adducts formed between 4-VP, 4-vinylguaiacol (4-VG) and 4-vinylcatechol (4-VC) with malvidin, peonidin or petunidin. These vinyl derivatives are decarboxylation products of the respective hydroxycinnamic acids. Similar observations were reported by Alcalde Eon et al. (2004) during fractionation and posterior identification of anthocyanins adducts synthesized in wine. Wang et al. (2003) identified a high percentage of anthocyanin adducts, the dominant pigments were hydroxyphenyl-pyranoanthocyanins formed by condensation with 4-VP and anthocyanins. In addition to the microbial enzyme activities, the release of free hydroxycinnamic acid can also be observed after the hydrolysis of the corresponding tartaric esters and, thus, contributes to the formation of pyranoanthocyanins.

Pinotin A is a kind of pyranoanthocyanin formed by the reaction of malvidin-3-glucoside and 4-VC. Nevertheless, Schwarz and Winterhalter (2003) reported the possible formation of this kind of pyranoanthocyanins by the direct addition of the anthocyanin with CA. Anthocyanin adduct resulting from the reaction with CA was also identified in Cabernet Sauvignon by Wang et al. (2003), however, products from the reaction with vinyl derivatives were found in higher concentration. Condensation of hydroxycinnamic acids and anthocyanins has been reported to proceed slower than products originated from vinyl derivatives (Wang et al., 2003; Schwarz et al., 2003a; de Freitas & Mateus, 2011).

### 2.3.2. Stability properties of pyranoanthocyanins

Pyranoanthocyanins are considered more stable than anthocyanins. They differ in many aspects compared to their anthocyanin precursors like chemical stability and the color intensity even at higher pH values (He et al., 2006).

Pyranoanthocyanins have a maximum absorption wavelength between 495 and 520 nm and a characteristic absorption maximum in the 420 nm region (Fulcrand et al., 1998; Morata et al., 2006; Marquez et al., 2013). Furthermore, the new pyran ring attached to the anthocyanin is responsible for the increased stability (Bakker & Timberlake, 1997; Fulcrand et al., 1998; Sousa et al., 2014). This addition provides more color stability due to the protection against nucleophilic attacks leading to colorless forms. The extension of the π electron system results in a hypsochromic shift of the absorption, responsible for the red-orange tonalities, while the positive charges given by the presence of oxygen atoms, enhance the resistance against oxidative degradation (Fulcrand et al., 1996; Sarni-Manchado et al., 1996; Schwarz & Winterhalter, 2003; Morata et al., 2003; Morata et al., 2012). At wine pH, this hypsochromic shift results in orange-brown colorations, while anthocyanins present more red colorations (Vivar-Quintana et al., 1999; Morata et al., 2006; Rentzsch et al., 2009).

The shift of equilibria to colorless forms is less favorable for pyranoanthocyanins compared to anthocyanins. While increasing the pH from 1 to 5 causes a loss of 80% of anthocyanins' color, pyranoanthocyanins hardly change their color intensity in this pH range (He et al., 2010a; de Freitas & Mateus, 2011). Similar results were reported by Vivar-Quintana et al. (1999). These authors studied the influence of pH and acetaldehyde in the formation of pyranoanthocyanins in microvinification under laboratory conditions. They reported a loss of anthocyanin content ranging from 65 to 70 % and attributed the presented color to the pyranoanthocyanins formed. Bakker and Timberlake (1997) conducted analysis of pyranoanthocyanins. These authors evaluated the effect of pH and color expression of different pyranoanthocyanins. At different pH-values pyranoanthocyanins exhibited an increased stability compared to their anthocyanins precursors (Mateus & de Freitas, 2001). In addition, these pigments displayed greater color expression at pH near neutrality. While anthocyanins undergo hydration reactions, pyranoanthocyanins undergo deprotonation reactions at varying pH (Cruz et al., 2010; Oliveira et al., 2013; Vallverdú-Queralt et al., 2016).

Copigmentation is believed to play an important role in the formation of pyranoanthocyanins (Fulcrand et al., 1998). For the food industry fortification of products with addition of extracts rich in copigments, i.e., hydroxycinnamic acids, could enhance anthocyanins stability (Maccarone at al., 1985; Talcott et al., 2003; Rein and Heinonen, 2004; Kopjar et al., 2011; Martillanes et al., 2017). During storage, anthocyanins can react with these compounds and form more stable compounds. As reported by Rein et al. (2005), the identified pyranoanthocyanins formed in strawberry and raspberry juices are formed with hydroxycinnamic acids, i.e., FA and sinapic acid.

### 2.3.3. Pyranoanthocyanins as food colorants

The demand for replacing synthetic food additives has aroused interest in finding natural sources. While anthocyanins are easily degraded, pyranoanthocyanins are proposed as candidates for use as food colorants, showing more resistance to temperature, pH variations, and stability during storage. Since pyranoanthocyanins have been reported to be products of anthocyanins reacting with several other compounds, the future might bring further identification of new members in this family of pigments. Nevertheless, none of the pyranoanthocyanins characterized up to now have been isolated in sufficient quantities to be used as a food colorant. After carrying out studies on *p*-hydroxyphenyl-pyranoanthocyanins, Vallverdú-Queralt et al. (2016) suggested that a wide range of colors can be obtained when the matrices are controlled well.

## 2.4. Hydroxycinnamic acids

Polyphenols include two major groups: phenolic acids and flavonoids. Phenolic acids are a family of phenolic compounds that possess one phenol ring bearing one or more hydroxy groups. They are widely distributed in fruits and vegetables (Clifford, 1999). Phenolic acids in plants act as protection against UV, insects, viruses, and microorganisms and are reported to contribute to flavor (Scalbert & Williamson, 2000; Manach et al., 2004).

Phenolic acids are a product of the shikimate pathway. Through this pathway and several enzymatic steps, intermediates of the pentose phosphate pathway and glycolysis phenylalanine and tyrosine are produced. These amino acids act as precursors for the formation of the phenolic acids (Taofiq et al., 2017). This family can be subdivided into two groups, namely hydroxybenzoic acids and hydroxycinnamic acids (HCA) (Figure 2.7). Hydroxycinnamic acids like are *p*-CA, FA and CA are among the most abundant compounds in nature. *p*-CA is an intermediate for the formation of CA, CQA, and FA (Taofiq et al., 2017).

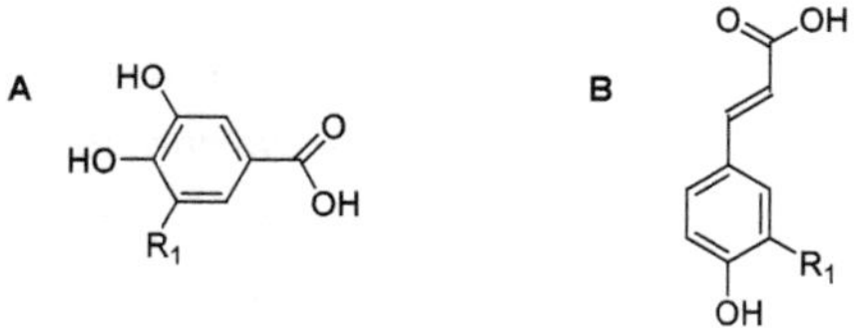

A

$R_1$= H: protocatechuic acid
$R_1$= OH: gallic acid

B

$R_1$= OH: caffeic acid
$R_1$= H: *p*-coumaric acid
$R_1$= $OCH_3$: ferulic acid

**Figure 2.7.** Structures of some hydroxybenzoic (A) and hydroxycinnamic acids (B)

HCA are rarely found in their free form but mostly as derivatives like amides combined with amino acids, peptides or as esters in combination with hydroxy acids or glycosides (Herrmann & Nagel, 1989; Jiang & Peterson, 2010; Taofiq et al., 2017).

Chlorogenic acid (CQA) is probably the most common HCA derivative. Important sources of chlorogenic acids are coffee (Clifford, 2000b; Clifford et al., 2006), apples (Schieber et al., 2001), potato (Dao & Friedman, 1992), and cacao, among others. CQA are esters of HCA.

The most common chlorogenic acid found in food is formed with CA and quinic acid (QA) known as 5-chlorogenic acid (5-CQA). Figure 2.8 shows the structures of these compounds.

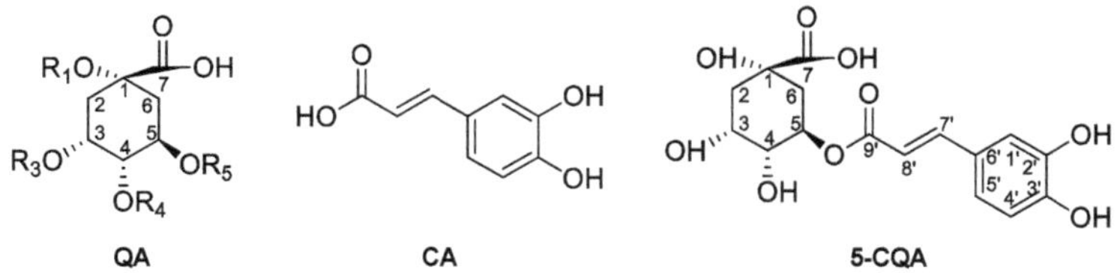

QA CA 5-CQA

**Figure 2.8.** Structures of quinic acid (QA), caffeic acid (CA) and as example of CQA 5-CQA

HCA are very reactive species. During thermal processing, HCA can undergo decarboxylation and oxidative reactions (Jiang & Peterson, 2010). These reactions have a significant impact on the flavor of the product. Bitter flavors also reported as "smoky", and "medicinal" flavors are attributable to naturally occurring phenolic acids from edible plants. Decarboxylation products were identified as major products of thermal reactions of FA. Decarboxylation reaction is catalyzed in the presence of oxygen yielding 4-VG. Further reactions could lead to the formation of volatile derivatives like 4-methyl and 4-ethylguaiacol (4-EG) from 4-VG. In the presence of oxygen, compounds, e.g., vanillin, acetovanillone and vanillic acid can be formed (Fiddler et al., 1967; Hua et al., 2013; Sun et al., 2018). In addition, HCA are probably the major contributors to the generation of undesirable aromas, also known as "horsy" and "leather" (Schieberle, 1991; Cabrita et al., 2012). These aromas are formed by the development of ethyl derivatives. Several authors have investigated the formation of aromas during thermal processing of popcorn (Schieberle, 1991); tortillas (Buttery & Ling, 1995), roasted peanuts (Lee et al., 1981), rice cakes (Buttery et al., 1999), coffee beans (Toci & Farah, 2008), and wine (Chatonnet et al., 1992; Chatonnet et al., 1995; Cabrita et al., 2006; Cabrita et al., 2012). Nevertheless, in wine, ethyl compounds are mainly produced by the enzymatic activity of microorganisms involved in the fermentation process. Undesired horsy odors are associated with the presence of ethyl compounds (Marín et al., 2005). Activities like cleavage of the ester bond, decarboxylation, and reduction of phenolic compounds are catalyzed by different microbial enzymatic activities. These metabolites can further react to the formation of more stable compounds. Among the number of microorganism capable of carrying out these activities, *Lactobacillus* have been extensively studied (Cavin et al., 1993; Couto et al., 2006; Curiel et al., 2010; Esteban-Torres et al., 2015).

### 2.5. Genus *Lactobacillus*

Lactic acid bacteria (LAB) are gram-positive, catalase-negative and non-spore forming microorganisms. They can occur as coccobacilli or as rods belonging to the Firmicutes (Felis & Dellaglio, 2007).

These microorganisms can be found in rich sources of carbohydrates, like food and plants. LAB are named after lactic acid, the main product of carbohydrate fermentation.

LAB are chemoautotrophic microorganisms, i.e., they can use organic and inorganic compounds for the production of energy. Regarding their carbohydrate metabolism, they can be classified as obligate homofermentative, facultative heterofermentative, or obligate heterofermentative. Obligate homofermentative LAB are able to ferment exclusively hexoses to lactic acid by the Emdben-Meyerhof-Parnas pathway. Facultative heterofermentative LAB metabolize pentoses and hexoses to lactic acid or acetic acid, while obligate heterofermentative LAB ferment hexoses and pentoses by the phosphogluconate pathway to lactic acid, ethanol or acetic acid, and carbon dioxide ($CO_2$).

In the food industry, the use of LAB as starter cultures for fermentation of vegetables and fruits is very common. They are used because they can easily be cultivated and are quite adaptive to different environments. They have also been shown to be very tolerant to high levels of phenolic compounds (Bel-Rhlid et al., 2013; Fritsch et al., 2017; Septembre-Malaterre et al., 2017). Besides their application in foods, some LAB have also been identified within the human microbiota. In recent years, an increasing number of *in vivo* studies have been published regarding the absorption of CQA and CA and their potential health benefits (Kono et al., 1995; Rice-Evans et al., 1996; Plumb et al., 1999; Sawa et al., 1999; Olthof et al., 2001). Several studies have been carried out to clarify the conversion pathways of phenolic acids by microorganisms (Bel-Rhlid et al., 2013; Filannino et al., 2015; Filannino et al., 2018).

### 2.5.1. Transformation of phenolic acid by Lactobacilli

Fermentation processes were already applied in ancient times primarily for food conservation. Beyond this purpose, the use of fermentation allows the creation of novel products. Fermentation can be defined as the result of controlled enzymatic activities of microorganisms on food substrates to transform them into more attractive products for consumers (Steinkraus, 1997; Marco et al., 2017; Filannino et al., 2018). Transformation of phenolic compounds by LAB has been reported to be species-specific, strain-specific (Sánchez-Maldonado et al., 2011; Filannino et al., 2018), as well as concentration-dependent. Some LAB are able to hydrolyze cinnamate esters into free HCA when they possess cinnamoyl esterase enzymes. Furthermore, LAB can also carry metabolic pathways for the degradation of phenolic acids such as decarboxylation and/or reduction. *L. reuteri*, *L. fermentum* and *L. helveticus* have been reported to be able to metabolize phenolic acids through degradation of ester bonds. Sánchez-Maldonado et al. (2011) reported a high concentration of CA as the product of the hydrolysis of CQA by *L. reuteri*, *L. fermentum* and *L. helveticus*. Phenolic acid decarboxylase enzymes with cinnamate decarboxylase activity have been reported in different groups of microorganisms, such as yeasts (Chatonnet et al., 1992; Cabrita et al., 2012) and some LAB (Cavin et al., 1993; Couto et al., 2006; Curiel et al., 2010; Buron et al., 2011; Silva et al., 2011). Metabolism of hydroxycinnamic acids is taking place by CD enzymes, decarboxylate the hydroxycinnamic acid into the vinylphenol derivative e.g. 4-VP from *p*-CA or 4-VG from FA. The action of vinylphenol reductase (VPR) causes reduction to ethylphenol, like 4-ethylphenol (4-EP) or 4-EG as the final product of this metabolic route. Nevertheless, a second metabolic pathway for phenolic acids has been reported (Barthelmebs et al., 2000; Curiel et al., 2010; Buron et al., 2011). This pathway eventually yields ethyl derivatives through a reduction catalysis by a phenolic acid reductase (PAR) followed by a decarboxylation presumptively conducted by a hydroxyphenylpropionic decarboxylase (HPD). Figure 2.9 shows both metabolic pathways for the decarboxylation and/-or reduction of hydroxycinnamic acids.

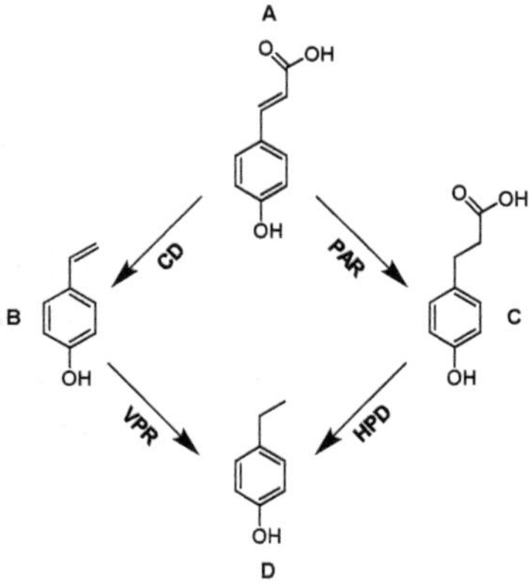

**Figure 2.9.** Pathways of LAB metabolism of *p*-CA (A) yielding 4-VP (B), by cinnamoyl decarboxylase activity (CD), PhA (C) by phenolic acid reductase (PAR), or 4-EP (D) by vinylphenol reductase (VPR) or hydroxyphenylpropionic decarboxylase (HPD) (modified after Buron et al., 2011)

The membrane of bacteria plays an important role since various functions are carried out there, i.e., regulation of substrate, metabolic products, and generation of energy. In addition, the membrane acts as a barrier to media preventing diffusion of extracellular compounds or cytoplasmatic material (Hutkins & Nannen, 1993; Konings et al., 1995;

Guerzoni et al., 2001). The membrane generates energy through the exchange of protons, electrons, and the generation of a proton motive force, also known as the electrochemical potential. Such proton motive force is necessary to achieve growth and replication at appropriate and not-appropriate conditions (Molenaar et al., 1993; Konings et al., 1995; Ribéreau-Gayon et al., 2006). The mode of action is based on an electron transport chain consisting of an asymmetrical distribution across the internal membrane with electric negative potential and positive proton gradient in the external part (ΔpH). In order to produce energy, NADH releases $H^+$ and two electrons which participate in the electron chain. This consists of transporting electrons from carrier to carrier through the membrane to the final acceptor, usually oxygen. The exchange of protons from inside to outside builds up a proton gradient ΔpH across the cell, translocating the negative charge inside the cell (Konings et al., 1995; Ribéreau-Gayon et al., 2006). This regulates the net consumption of protons from the cytoplasm and their release outside the cell. Therefore, affluence of protons promotes the synthesis of ATP, while the discharge of protons leads to a consumption of energy (Molenaar et al., 1993). The cell keeps an optimum pH in order to maintain all cellular functions. Nevertheless, the cell can induce changes in membrane composition to assure cell viability, as a response to external factors affecting cellular growth, i.e., temperature, exposure to toxic compounds or change on pH media (Hutkins & Nannen, 1993; Castro et al., 1996; Rozès & Peres, 1998; Fernández Murga et al., 2000; Ulmer et al., 2002; Tymczyszyn et al., 2005). Furthermore, a drop in the pH media provokes an increment in the electrical potential. These differences in charge between the cytoplasm and the external media cause a dissipation of the proton motive force which may lead to cell death (Aertsen & Michiels, 2004; Ribéreau-Gayon et al., 2006). The specific reason for the metabolism of phenolic acids by LAB is still unknown, but some hypotheses have been proposed. At acidic pH, phenolic acids are found in their protonated form. The dissipation of the ΔpH for possible interactions between the phenolic acids and the cell membrane allows their absorption due to a neutralization of the membrane electric potential. A toxic effect of phenolic acids has been proposed since they can easily enter the cell in their undissociated form causing damages to the membrane, finally leading to a leakage of the internal cellular constituents. It has been suggested that a detoxifying proton pump is activated in order to stop the dissipation of the membrane. This proton pump is pumping protons out of the cell to restore the ΔpH (Carmelo et al., 1997; Barthelmebs et al., 2000).

Couto et al. (2006) screened 35 LAB strains regarding their cinnamoyl decarboxylase activity in the presence of two hydroxycinnamic acids and the ability of a reduction step to ethylphenol conducted by few of them. Strains of *L. plantarum*, *L. mali,* and *L. sakei* were able to decarboxylate *p*-CA and FA exhibiting high decarboxylation rates and low reductase activities.

Svensson et al. (2010) reported esterase activities, phenolic decarboxylase and reductase activities, as well as glucosidase activities during fermentation of sorghum doughs by *L. fermentum*, *L. casei*, *L. reuteri,* and *L. plantarum*. These authors reported that Lactobacillus strains use different pathways to metabolize phenolic acids and flavonoids. The products released after phenolic acid degradation have been identified as precursors in the formation of pyranoanthocyanins (Schwarz et al., 2003a; Schwarz et al., 2004; Hillebrand et al., 2004; Bai et al., 2014). In addition, other products resulting from microbial activity can be found the production of biogenic amines, formed after decarboxylation of amino acids or by amination of and transamination of aldehydes and ketones. These biogenic amines can be found in raw and processed food e.g. fermented food. Interests in these compounds are mainly due to health risk concerns. In winery, LAB are the main agent for the formation of these compounds (Herbert et al., 2005; Landete et al., 2005; Ali et al., 2010).

## 2.6. Black carrot juice

Cultivated carrots are divided into two groups: the western or carotene group which is yellow, orange and sometimes white (*Daucus carota* ssp. *sativus* var. *sativus*) and the eastern or anthocyanin group with purple, pink orange-yellow hues (*Daucus carota* ssp. *sativus* var. *atrorubens* Alef.) (Stolarcyk & Janick, 2011). They have been cultivated for at least 3 000 years. Black carrots are still largely unknown in the western world although they are consumed in countries such as Turkey, Afghanistan, Egypt, Pakistan, India, and in the Far East.

Black carrots are rich sources of anthocyanins and phenolic acids, alkaloids, terpenoids, coumarin, among other bioactive compounds. Black carrot anthocyanidins are primarily cyanidin derivatives, but traces of pelargonidin and peonidin have also been identified (Malien-Aubert et al., 2001; Kammerer et al., 2003; Schwarz et al., 2004; Montilla et al., 2011). Kammerer et al. (2004a) characterized black carrots and reported hydroxycinnamic acids derivatives as the major constituents in black carrots. Regarding hydroxycinnamic acids, 5-CQA have been reported as the main phenolic acid (Kammerer et al., 2004a; Esatbeyoglu et al., 2016). Black carrot anthocyanins were found acylated with hydroxybenzoic or hydroxycinnamic acids, e.g., CA, FA, or *p*-CA (Malien-Aubert et al., 2001; Kammerer et al., 2003, 2004a, 2004b; Arscott & Tanumihardjo, 2010; Gras et al., 2016).

Black carrot anthocyanins are stable at higher pH values compared to anthocyanins from other edible plants. This higher stability has been reported for solutions with pH above 5, which is not expected for anthocyanins. This stability has been attributed to the occurring acylation. As previously described, acylated anthocyanins exhibit greater color stability compared with their non-acylated analogs. This improvement in their stability is mainly due to intramolecular interactions making them more resistant to hydration (Malien-Aubert et al., 2001; Schwarz & Winterhalter, 2003; Gras et al., 2016). Kammerer et al. (2004b) reported black carrots from different cultivars possessing acylations in a range from 55 % to 99 %. In agreement with these authors, Stintzing et al. (2002) reported 41 % of anthocyanins presenting acylations. The stability presented by black carrot anthocyanins makes them good candidates for their use in the food industry where they can be applied under the E number 163. The commercial use of black carrot anthocyanins in products such as jellies, confectionery, soft drinks or some conserves was introduced by Oversal Foods Ldt. under the name of "Carantho". In addition, the application of black carrot concentrates is also possible as coloring food stuff, whereby the declaration does not need an E number. Therefore, using black carrots as coloring food stuff represents an attractive alternative for food coloring.

## 3. AIM OF THE STUDY

The objective of the present study was to formulate a juice rich in pyranoanthocyanins formed during fermentation of black carrot juice and to produce an applicable coloring foodstuff.

This work had three main objectives:

- To evaluate the capacity of LAB to produce vinyl derivatives from hydroxycinnamic acids by hydroxycinnamate decarboxylase activity.
- To evaluate the specificity of cinnamoyl esterase activity displayed by Lactobacillus species in the presence of monochlorogenic and dichlorogenic acids.
- To synthesize hydroxyphenyl-pyranoanthocyanin through providing their precursors by fermentation of black carrot juice by LAB

# 4. MATERIALS AND METHODS

## 4.1. Material

### 4.1.1. Plant material

For the present work, blackberry concentrate (69.9 °Brix) was kindly provided by Wild (Heidelberg, Germany) and kept at -21 °C until anthocyanin isolation. Locally (Bonn, Germany) purchased black carrots (*Daucus carota* ssp. *sativus* var *atrorubens* Alef.) cultivars beta sweet and deep purple black carrots were stored at -21 °C until further processing. Jerusalem artichokes (*Helianthus tuberosus* L.) were purchased from a local market in Bonn (Germany), and kept at -80 °C until use. Green Robusta coffee beans (*Coffea canephora* Pierre ex A. Froehner) were kindly provided by the Institute of Food Chemistry, Technische Universität Braunschweig (Germany), and were stored at -80 °C until further use.

### 4.1.2. Microbiological material

The bacteria strains used in this work, were obtained from Leibniz Institute DSMZ. Lactic acid bacteria were selected on the basis of their enzymatic activities. Microorganism and their enzymatic activities are listed in table 4.1.

**Table 4.1.** Microorganism used in microbiological experiments

| Microorganisms | Strain | Enzymatic activity |
|---|---|---|
| *L. mali* | DSM 20444 | Cinnamate decarboxylase |
| *L. plantarum* | DSM 20174 | Cinnamate decarboxylase |
| *L. sakei subsp. sakei* | DSM 20017 | Cinnamate decarboxylase |
| *L. reuteri* | DSM 20016 | Esterase cinnamoyl |
| *L. helveticus* | DSM 20075 | Esterase cinnamoyl |
| *L. fermentum* | DSM 20052 | Esterase cinnamoyl |
| *Sacchrothix espanaensis* | DSM 44229 | Tyrosine ammonia lyase |

### 4.1.3. Chemicals

Table 4.2 and 4.3 show the list of chemicals for analytical and microbiological experiments and the corresponding suppliers used in this work. All reagents and solvents needed for analyses were of HPLC, MS or analytical grade.

**Table 4.2.** Chemicals used for analytical analysis

| Substance | Purity | Supplier |
|---|---|---|
| Acetic acid | ≥ 99.5 % | VWR International GmbH, Darmstadt, Germany |
| Acetonitrile | LC-MS | Th. Geyer GmbH & Co. KG Renningen, Germany |
| Amberlite® XAD7-HP | | Sigma Aldrich |
| Caffeic acid | ≥ 95% | Sigma Aldrich, Steinheim, Germany |
| Chlorogenic acid | ≥ 95 % | Sigma Aldrich, Steinheim, Germany |
| Ethanol | 99.8 % | Carl Roth GmbH + Co. KG, Karlsruhe |
| Ethyl acetate | ≥ 99.5 % | Fischer chemical, Loughborough, UK |
| Ethylphenol | ≥ 99.9 % | Sigma-Aldrich, Steinheim, Germany |
| Ferulic acid | ≥ 99.9 % | Sigma-Aldrich, Steinheim, Germany |
| Formic acid | LC-MS | Sigma Aldrich, Steinheim, Germany |
| Hydrochloric acid | 37 % | Carl Roth, Karlsruhe, Germany |
| Lactic acid | | Fluka, Sigma Aldrich Chemie GmbH, Spain |
| Methanol | HPLC Grade | Th. Geyer GmbH & Co. KG, Renningen, Germany |
| n-Butanol | HPLC Grade | Fisher Scientific, Loughborough, UK |
| *p*-coumaric acid | | Sigma-Aldrich, St. Louis, MO, USA |
| Sodium Chloride | | Sigma-Aldrich, St. Louis, MO, USA |
| Sodium hydroxide | | VWR, Darmstadt, Deutschland |
| 4-Vinylphenol | 10 wt % in Propylene glycol | Sigma Aldrich, St. Louis, MO, USA |
| 4-Vinylguaiacol | ≥ 98.0 % | Sigma Aldrich, St. Louis, MO, USA |
| Water | Bi-distilled | Th. Geyer GmbH & Co. KG, Renningen, Germany |
| Water | | PURELAB Flex 2 Veolia Water Technologies Deutschland, Celle, Germany |

**Table 4.3.** Chemicals used for media preparation

| Substance | Supplier |
|---|---|
| Ammonium citrate tribasic | Sigma Aldrich, Steinheim, Germany |
| L-Arginine | Sigma Aldrich, St. Louis, MO, USA |
| Ascorbic acid | Fluka Analytical, Steinheim, Germany |
| L-Cysteine | Sigma Aldrich, St. Louis, MO, USA |
| Dipotassium phosphate | Merck, Darmstadt, Germany |
| Glucose | Carl Roth, Karlsruhe, Germany |
| L-Glutamate | Fluka Analytical, Steinheim, Germany |
| L-Isoleucine | Sigma Aldrich, St. Louis, MO, USA |
| L-Leucine | Sigma Aldrich, St. Louis, MO, USA |
| Magnesium chloride | Merck, Darmstadt, Germany |
| Magnesium dichloride | Merck, Darmstadt, Germany |
| Magnesium sulphate | Merck, Darmstadt, Germany |
| Manganese sulfate | Merck, Darmstadt, Germany |
| Malt extract | Extrakt Chemie, Stadthagen, Germany |
| Meat extract | Carl Roth, Karlsruhe, Germany |
| L-Methionine | Sigma Aldrich, St. Louis, MO, USA |
| Niacin | Sigma Aldrich, St. Louis, MO, USA |
| Phantothenic acid | Carl Roth, Karlsruhe, Germany |
| L-Phenylalanine | Carl Roth, Karlsruhe, Germany |
| Peptone | Fluka Analytical, Steinheim, Germany |
| Pyridoxymine | Acros Organics, Geel, Belgium |
| Riboflavin | Fluka Analytical, Steinheim, Germany |
| Sodium acetate | Carl Roth; Karlsruhe, Germany |
| L-Tryptophan | Fluka Analytical, Steinheim, Germany |
| Tween 80 | Carl Roth; Karlsruhe; Germany |
| L-Tyrosine | Sigma Aldrich, St. Louis, MO, USA |
| L-Valine | Sigma Aldrich, St. Louis, MO, USA |
| Yeast extract | Leiber, Bramsche, Germany |

### 4.1.4. Machines and equipment

Identification and/or quantification of products were carried out by ultra-high-pressure liquid chromatography with diode array detection (UHPLC-DAD), and UHPLC-DAD coupled to electrospray ionization mass spectrometry (UHPLC-ESI-$MS^n$). Machine specifications are listed in Tables 4.4 and 4.5, respectively.

**Table 4.4.** Specifications of the UHPLC-DAD

| UHPLC-DAD | |
|---|---|
| UHPLC | Schimadzu Prominence System, Duisburg, Germany |
| Pump | Nexera X2 LC-30AD |
| Degasser | DGU-20A$_{5R}$ |
| Autosampler | SIL-30AD |
| Pre-column | Acquity UPLC HSST3 (2.1 x 5 mm, 1.8 µm), Waters, Milford, USA |
| Column | Acquity UPLC HSST3 (150 x 2.1 mm, 1.8 µm), Waters, Milford, USA |
| Column oven | CTO-20AC |
| PDA Detector | SPD-M20A |

**Table 4.5.** Specifications of the UHPLC-ESI-MS$^n$

| UHPLC-ESI-MS$^n$ | |
|---|---|
| UHPLC | Waters Acquity I-Class system, Milford, USA |
| Mass spectrometer | LTQ-XL, Thermo Scientific Inc., Braunschweig, Germany |
| Ionization | Electrospray-Interface (ESI) |
| Pump | Binary system |
| Detector | DAD |
| Pre-column | Acquity UPLC HSS T3 VanGuard (2.1 x 5 mm, 1.8 µm) |
| Column | Acquity UPLC HSS T3 (150 x 2.1 mm, 1.8 µm) Waters Corporation, Milford, USA |
| Software | ThermoXcalibur (Version 2.2 SP 1.48) |

The separations of chlorogenic acid fractions were performed by high-performance countercurrent chromatography (HPCCC). Specifications of the machine is listed in Tables 4.6 and 4.7.

**Table 4.6.** Specifications of the HPCCC

| HPCCC | |
|---|---|
| HPCCC | Dynamic Extractions Ldt., Berkschire, UK |
| Pump | HPLC K-501, Knauer, Berlin, Germany |
| Detector | K-2500, Knauer, Berlin, Germany |
| Fraction collector | LKB super Frac., Pharmacia, Uppsala, Sweden |

**Table 4.7.** Specification of the semi-preparative HPLC

| Semi-preparative HPLC | |
|---|---|
| HPLC | Knauer, Berlin, Germany |
| Pump | 1050, Teledyne ISCO, Lincoln, USA |
| Autosampler | 3950, Teledyne ISCO, Lincoln, USA |
| Manager | 5050, Teledyne ISCO, Lincoln, USA |
| Column | Eurospher II 100-5C18 (5 µm, 250 x 16 mm), Teledyne ISCO, Lincoln, USA |
| Detector | 4 channel UV/vis 2550, Teledyne ISCO, Lincoln, USA |
| Fraction collector | Foxy R1, Teledyne ISCO, Lincoln, USA |
| Software | EZChrom Elite |

Table 4.8 lists further equipment used for the realization of this work.

**Table 4.8.** Equipment

| Machine | Model | Supplier |
|---|---|---|
| Analytical balance | AB304-S/FACT | Mettler Toledo, Columbus, OH, USA |
| Autoclave | Sanoclav LAM-4-20-MCS-J | Wolf; Bad Überkingen, Germany |
| Freezer (-80 °C) | Forma™ 900 | Thermo Fisher Scientific, USA |
| Cooled incubator | Cooled incubators Type 3001 | Rubarth Apparate, Laatzen, Germany |
| Laboratory shaker | Orbital shaker SSL1 | Stuart, Stone, UK |
| Refrigerator | FKS 50001 | Liebherr, Bulle, Switzerland |
| Microcentrifuge | Heraeus™ Pico 17 | Thermo Scientific, Watham, MA, USA |
| pH-meter | pH 3210 WTW | WTW, Weilheim, Germany |
| pH-meter | pH-meter 765 | Knick, Berlin, Germany |
| Spectrophotometer | Genesys 6 | Thermo Scientific, Watham, MA, USA |
| Laminar flow cabinet | HF72 | Gelaire Flow Laboratories, Australia |
| Ultrasonic bath | Ultrasonic cleaner | VWR, USA |
| Precision balance | PB1502-L | Mettler Toledo, Columbus, OH, USA |
| MS filters | Chromafil RC-20/15 | Manchery-Nagel, Düren, Germany |
| | Chroma Meter CR400 | Minolta Konica, Japan |
| Calibration plate | Calibration plate No. 2 CR-210/CR-310 | Minolta Konica, Japan |

| Machine | Model | Supplier |
|---|---|---|
| Cuvvetes | 692.104-BT | Minolta Konica, Japan |
| Membrane adsorber | Sartobind® S IEX 150 mL | VWR, Darmstadt, Deutschland |
| Membrane adsorber pre-filter | Sartopore® 2 300 | VWR, Darmstadt, Deutschland |
| Pump | Pumpdrive 5201 | Heidolph |
| Ultracentrifugal mill | ZM 100 | Wilhelm Werner GmbH., Leverkusen, Germany |
| Juice extractor | MESOAO 77 W | Bosch, China |

## 4.2. Preparative methods

### 4.2.1. Preparation of culture media and buffer solutions

#### 4.2.1.1. Media

Media de Man Rugosa and Sharpe (MRS) was used for cultivation of all lactobacilli, whereas Glucose, Yeast and Malt (GYM) media was used for cultivation of *S. espanaensis*. In the following Tables 4.9, 4.10, and 4.11, the compounds needed for preparation of media used in this work are listed.

**Table 4.9.** Constituents of MRS media (pH 6.6)

| Compound | Concentration (g $L^{-1}$) |
|---|---|
| Peptone | 10 |
| Yeast extract | 4 |
| Tween 80 | 1 |
| Sodium acetate | 5 |
| Magnesium sulphate | 0.2 |
| Beef extract | 8 |
| Glucose | 20 |
| Dipotassium phosphate | 2 |
| Ammonium citrate tribasic | 2 |
| Manganese sulfate | 0.05 |

**Table 4.10.** Composition of GYM media (pH 7.2)

| Compound | Concentration (g $L^{-1}$) |
|---|---|
| Glucose | 4 |
| Yeast extract | 4 |
| Malt extract | 10 |

**Table 4.11.** Composition of PMM7 media (pH 6.6)

| Compound | Concentration (g $L^{-1}$) |
|---|---|
| Dipotassium phosphate | 8.6 |
| Sodium acetate | 1 |
| Ammonium citrate tribasic | 0.6 |
| Magnesium chloride | 0.2 |
| Manganese dichloride | 0.16 |

| Compound | Concentration (g $L^{-1}$) |
|---|---|
| Ascorbic acid | 0.5 |
| Arginine | 0.125 |
| Cysteine | 0.130 |
| Glutamate | 0.5 |
| Isoleucine | 0.210 |
| Leucine | 0.475 |
| Methionine | 0.125 |
| Phenylalanine | 0.275 |
| Tryptophan | 0.225 |
| Tyrosine | 0.250 |
| Valine | 0.325 |
| Niacin | 0.001 |
| Phantothenic acid | 0.001 |
| Pyridoxymine | 0.005 |
| Riboflavin | 0.001 |
| Glucose | 11 |

#### 4.2.1.2. Buffers

Phosphate buffer consisted of a mixture of 1 M of disodium hydrogen phosphate ($Na_2HPO_4$ 22.415 g $L^{-1}$) and 1 M of potassium dihydrogen phosphate ($KH_2PO_4$ 9.073 g $L^{-1}$). By varying the amount of each solution, different pH values can be adjusted (Table 4.12).

**Table 4.12.** Composition of phosphate buffer (100 mL)

| pH | $Na_2HPO_4$ (mL) | $KH_2PO_4$ (mL) |
|---|---|---|
| 6.0 | 12.00 | 88.00 |
| 6.2 | 18.5 | 81.5 |
| 6.4 | 26.5 | 73.5 |
| 6.6 | 37.5 | 62.5 |

For preparation of McIlvaine buffer, two solutions are needed, 0.2 M stock solution of disodium phosphate ($Na_2HPO_4$ $2H_2O$ 35.60 g $L^{-1}$) and 0.1 M stock solution of citric acid (citric acid 21.01 g $L^{-1}$). McIlvaine buffer can be prepared as indicated in Table 4.13.

**Table 4.13.** McIlvaine buffer composition (100 mL)

| pH | $Na_2HPO_4$ (mL) | Citric acid (mL) |
|---|---|---|
| 4.6 | 53.25 | 46.75 |
| 4.8 | 50.70 | 49.30 |
| 5.0 | 48.50 | 51.50 |
| 5.2 | 46.40 | 53.60 |
| 5.4 | 44.25 | 55.75 |
| 5.6 | 42.00 | 58.00 |
| 5.8 | 39.55 | 60.45 |
| 6.0 | 36.85 | 63.15 |
| 6.2 | 33.90 | 66.10 |
| 6.4 | 30.75 | 69.25 |
| 6.6 | 27.25 | 72.75 |

### 4.2.2. Processing of blackberry juice

Blackberry juice was initially cleaned up using column chromatography on Amberlite XAD-7 HP porous polymer. A column (64 x 10 cm) filled with Amberlite XAD-7 HP was equilibrated with water until the washing solution was completely clear. The extract was obtained by dilution of blackberry concentrate with distilled water (1:10, v:v) and applied onto the column. For the removal of sugars, proteins, and organic acids, the column was rinsed with 10 L distilled water and, subsequently, a polyphenolic extract was eluted with EtOH/AcOH (19:1, v:v). After elution, the solvent was removed using a rotary evaporator and subsequently, the polyphenolic extract was lyophilized.

### 4.2.3. Fractionation of anthocyanins and anthocyanin adducts by membrane adsorber

For isolation of anthocyanins and pyranoanthocyanins ion exchange membrane chromatography was applied. Isolation was carried out by Sartobind membrane adsorber protected by a Sartopore pre-filter and connected to a pump for delivering the solvents. 3 g of the polyphenolic extract (section 4.2.2) were brought to a final volume of 1 L with distilled water. The solution was filtered prior to loading on the membrane adsorber.

Before separation, the membrane adsorber was regenerated and equilibrated with 2.5 L 1 M of NaOH and 2.5 L 1 M HCl, then 1 L of EtOH/AcOH (19:1; v:v) was loaded to the membrane adsorber for conditioning before applying the polyphenolic extract. Fractionation of anthocyanins was carried out in two steps. First, the membrane was washed with 2.5 L of EtOH/AcOH (19:1; v:v) to remove copigments and, in the second step, 1 M NaCl solution mixed with EtOH in a ratio 1:1 (v:v) was applied for the elution of anthocyanins. The stabilization of anthocyanins stabilization was done by acidifying the elution solution with 1 % of the total volume with acetic acid. A 1 M aqueous NaCl solution 2:8 (v:v) with EtOH was used to store the membrane. Anthocyanins were further purified (see section 4.2.2), subsequently lyophilized and stored at -25 °C (figure 4.1).

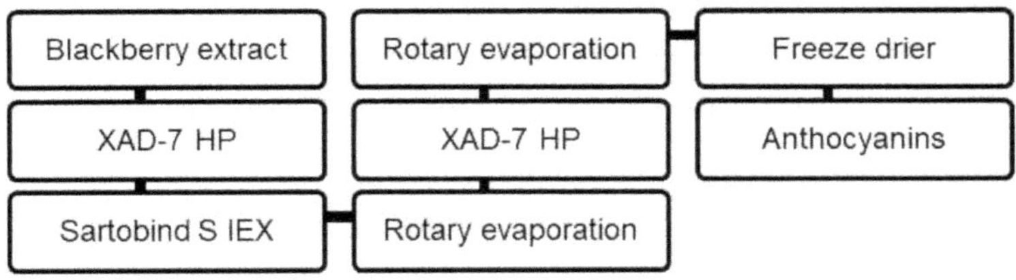


**Figure 4.1.** Fractionation of blackberry anthocyanins

### 4.2.4. Preparation of chlorogenic acid extracts

Before extraction of CQA, tubers of Jerusalem artichokes and green coffee beans were separately lyophilized and subsequently ground with an ultracentrifugal mill ZM 100 to a particle size of 0.25 mm. Green coffee beans powder (0.2 g) was extracted twice with 25 mL of aqueous methanol solution (1:1, v:v) at 60 °C for 15 min in an ultrasonic bath. Samples were then centrifuged for 5 min at 8 000 rpm and the pellet was subjected to a second extraction under the same conditions. Extracts were evaporated under vacuum at 35 °C to remove methanol, lyophilized, and kept at -80 °C until HPCCC separation (section 4.2.5). The same procedure was applied to powder of Jerusalem artichoke for preparation of CQA extracts.

### 4.2.5. Chlorogenic acid isolation by HPCCC

The separations of CQA fractions were performed with a Spectrum HPCCC instrument from Dynamic Extractions Ltd. The UV- absorbance was monitored at 324 nm. Manual loop injection (5 mL) was used. The solvent system for separation was ethanol/n-butanol/acetic acid/n-hexane/water (2:2:0.5:1:5, v/v). An aliquot of 0.5 g of green coffee beans or Jerusalem artichoke extract was dissolved in a mixture 1:1 (v:v) of both HPCCC eluents. The separations were conducted using the lower dense phase as the stationary phase and then the apparatus was rotated at 1500 rpm and a flow rate of 4 mL $min^{-1}$. The fractions were collected in 2 min intervals in test tubes with a fraction collector. The separation method allowed the recovery of six fractions for green coffee beans samples, whereby some fractions contained compounds in high purity. The fractions containing 4-CQA and 5-CQA was subjected to semi-preparative HPLC for purification of 4-CQA. During fractionation of both green coffee beans and Jerusalem artichoke extracts, di-CQAs remained in the coil. This fraction was directly used for the evaluation of enzyme activity regarding cleavage of di-CQAs.

### 4.2.6. Separation of 4-CQA by semi-preparative HPLC

The HPCCC fraction containing the isomers of 4-CQA and 5-CQA acid, was separated by semi-preparative HPLC. Therefore, the sample (12 mg $mL^{-1}$) was dissolved in 0.1% formic acid in water (v $v^{-1}$) and filtered. Solvent A consisted of 0.1 % formic acid in water (v $v^{-1}$) while acetonitrile was used as solvent B. For separation, the gradient program shown in table 5.14 was applied. The flow rate was set at 6 mL $min^{-1}$, the injection volume was 600 µL, and the separation was monitored at a wavelength of 324 nm. Fractions were collected and afterwards, the solvent was removed using lyophilization. Isolated compounds were stored at -80 °C until further use.

**Table 5.14.** Gradient used for semi-preparative UHPLC separation of fractions of 4-CQA and 5-CQA

| Time (min) | Eluent B (%) |
|---|---|
| 0 | 12 |
| 12 | 25 |
| 25 | 35 |
| 30 | 100 |
| 35 | 100 |
| 40 | 45 |
| 45 | 12 |

### 4.2.7. Identification of black carrot juice anthocyanins

Black carrots were washed, peeled and blended using a mechanical juice extractor. The resulting black carrot juice (BCJ) was filled into bottles and pasteurized at 90 °C for 15 min in a hot water bath. The BCJ was stored at -80 °C. Anthocyanin content was quantified using Cy-3-glu as an external standard (section 4.4.3 and Table 4.18) and a molar mass correction factor was applied.

## 4.3. Microbiological methods

### 4.3.1. Decarboxylation of free hydroxycinnamic acids by Lactobacilli

*L. plantarum*, *L. mali* and *L. sakei* were grown in MRS media at 35 °C until the culture reached the late exponential phase. McIlvaine's buffer and MRS media were supplemented separately with *p*-CA, CA, or FA. Experiments were conducted at pH values of 4.6 and 5.6. Before sterilization (121 °C, 15 min), the pH was adjusted with 6 M HCL. Stock solutions of each HCA were prepared and added after sterilization. Hydroxycinnamic acids were supplemented to the media to get a final concentration of 3.05 mM for *p*-CA, 2.77 mM for CA or 2.57 mM for FA. The inoculum consisted of 5 % of cultures in the late exponential phase, they were washed twice with a sterile saline solution (0.9 % NaCl) and re-suspended in MRS media or buffer each containing phenolic acid. Buffer solution and MRS media, each at pH values in the presence of the three HCA without inoculum, served as a control. Lactic acid fermentation was done at 35 °C in a rotary shaker (150 rpm) for 24 h and each experiment was performed in duplicate. After 24 h, a 1 mL sample was taken, centrifuged at 10 000 rpm for 5 min and the supernatants was membrane filtered UHPLC-DAD analysis (section 4.4.2 and table 4.15).

### 4.3.2. Metabolism of phenylpropanoid derivatives involved in the synthesis of *p*-CA

*L. plantarum*, *L. mali* and *L sakei* were screened for their ability to metabolize different compounds involved in the phenylpropanoid pathway. MRS media (pH 6.6) was supplemented separately with *p*-CA, 4-VP, 4-EP, PhA, CNA, L-Tyr or L-Phe at a concentration of 1 mM. For this purpose, fresh solutions were prepared and added to the growth media after sterilization. Assays were performed with 5 % inoculum of overnight cultures at 35 °C under shaking (150 rpm) for 24 h. Each individual assay was conducted in duplicate and control assays were prepared for all compounds without inoculum. After fermentation, samples were taken, centrifuged at 10 000 rpm for 5 min and the supernatant was membrane filtered for analysis performed by UHPLC-DAD (section 4.4.2 and Table 4.16).

### 4.3.3. Effect of *p*-coumaric acid concentration on the production of volatile derivatives in MRS media

Formation of 4-VP during fermentation of MRS media containing *p*-CA with *L. plantarum*, *L. mali* and *L. sakei* was examined. MRS media (pH 6.6) was supplemented after sterilization with *p*-CA at each final concentration of 0.1 mM, 0.2 mM, 0.5 mM, 1 mM, 2 mM, 3 mM or 5 mM. Media was inoculated with 5 % of overnight cultures of bacterial strains. Assays were conducted in duplicate, MRS medium at the same concentrations of *p*-CA without inoculum were prepared and used as control. Fermentation was carried out at 35 °C under shaking at 150 rpm for 18 h. After fermentation, 1 mL samples were taken, centrifuged and membrane filtered. Analysis of samples was performed by UHPLC-DAD applying the conditions described in section 4.4.2 and Table 4.16.

### 4.3.4. Decarboxylation of *p*-coumaric acid in minimal PMM7 media

Decarboxylation activities of *L. plantarum*, *L. mali* and *L. sakei* were studied in a minimal PMM7 media. Bacterial cultures were grown separately in MRS media at 35 °C overnight. An inoculum consisting of 5 % overnight cultures was taken, centrifuged at 9 000 rpm for 1 min, the pellet was recovered and washed twice with a sterile solution (0.9 % NaCl) and afterward re-suspended in the PMM7 media containing 0.2 mM, 2 M or 5 M of *p*-CA. Fermentations were conducted in duplicate at 35 °C under shaking at150 rpm for 48 h. Control assays were prepared in PMM7 media without inoculum. After fermentation, 1 mL of each fermentation was taken, centrifuged and membrane filtered. Analysis of samples was carried out by UHPLC (section 4.4.2). Table 4.16 shows the gradient applied for analysis.

### 4.3.5. TAL activity of *S. espanaensis* in GYM media

GYM media was supplemented with L-Tyr for TAL activity test. GYM media containing L-Tyr at a final concentration of 1 µM was inoculated with 2 mL of *S. espanaensis* culture. Experiment was carried out in duplicate at 28 °C under shaking at 150 rpm during 7 d. Parallel experiments were conducted in GYM media supplemented with *p*-CA at a final concentration of 2 mM in absence of L-Tyr for TAL activity test. The experiment was carried out in duplicate at 35 °C in a rotary shaker (150 rpm) for 72 h. The inoculum consisted in 5 % of overnight *L. plantarum* culture, washed twice with a sterile saline solution (0.9 % NaCl) and re-suspended in GYM media for fermentation. GYM media containing *p*-CA was used as control.

After fermentation, 1 mL samples were taken, centrifuged and membrane filtered. Samples were analyzed by UHPLC-DAD applying the conditions described in table 4.16 in section 4.4.2. Detection of products was performed from 225 nm to 320 nm.

### 4.3.6. Metabolism of 5-CQA during fermentation by LAB

*L. reuteri*, *L. helveticus* and *L. fermentum* were grown in MRS media at 35 °C. Before sterilization the pH value was adjusted to 6.6 with 6 M HCl. To initiate cinnamoyl esterase synthesis, the strains were grown in the presence of 5-CQA (300 µM) for 12 h.

To evaluate cinnamoyl esterase activity, a 10 mL inoculum containing approximately $10^8$ CFU $mL^{-1}$ of the overnight cultures was taken. Bacteria were centrifuged for 5 min at 8 000 rpm, washed twice with sterile saline solution (0.9 % NaCl) and finally re-suspended in MRS media containing 1 mM of 5-CQA. MRS media containing the same concentration of 5-CQA without inoculum was used as control. Fermentations were performed at 35 °C for 24 h under agitation (150 rpm). Samples were taken after 12 and 24 h, centrifuged and filtered for UHPLC analysis.

Quantification of products was carried out by UHPLC-DAD following the gradient specified in Table 4.15 (section 4.4.2).

### 4.3.7. Evaluation of cinnamoyl esterase activity of *L. reuteri*, *L. helveticus* and *L. fermentum*

*L. reuteri*, *L. helveticus*, and *L. fermentum* were grown in MRS media at 35 °C, for experiments an inoculum consisting of 10 mL containing approximately $10^8$ CFU $mL^{-1}$ of overnight cultures was used. Bacteria were centrifuged for 5 min at 8 000 rpm, washed twice with sterile saline solution (0.9 % NaCl) and were finally re-suspended in 1 mL of sterile water. All mono CQA isomer (3-CQA, 4-CQA and 5-CQA) or a mixture of di-CQAs were dissolved in water and filter sterilized. MRS media (pH 6.6) was supplemented with 1 mM of each mono CQA isomer or a mixture of di-CQAs after sterilization. The media was then inoculated with the cultures and incubated at 37 °C. Samples were taken at 0, 12, 24, 48, and 72 h for quantification of CQA and CA. For each CQA isomer, a control without addition of inoculum was conducted simultaneously. Controls were acidified to pH 4.5 to simulate the conditions which were detected in the fermented samples. Quantification of products was carried out by UHPLC-ESI-$MS^n$ (section 4.4.4).

### 4.3.8. Synthesis of hydroxyphenyl-pyranoanthocyanins in buffer and MRS media

The fermentation products of *L. sakei* from decarboxylation activities (section 4.3.1) were centrifuged at 10 000 rpm for 5 min, the pellet was removed and media was supplemented with 1.78 mM Cy-3-glu. The reaction was performed for 5 d, afterwards samples were centrifuged at 10 000 rpm for 5 min and the supernatant was membrane filtered for quantification by UHPLC-DAD. Analysis of samples were performed following the gradient shown in Table 4.18 in section 4.4.3.

### 4.3.9. Stability of Cy-3-glu and pyranoanthocyanins derivatives at fermentation conditions

Samples of the fermentation performed according to section 4.3.8 were recovered and centrifuged at 10 000 rpm for 10 min. The supernatant was collected and cleaned-up using column chromatography on Amberlite XAD-7 HP to remove residues of the MRS media. Procedure is detailed in section 4.2.2. After washing, pyranoanthocyanin products of the fermentation were isolated by ion exchange membrane chromatography (section 4.2.3). These isolated pyranoanthocyanins and their precursor, i.e., Cy-3-glu were used for evaluation of their stability simulating fermentation conditions. Evaluation was carried out at pH values of 4.6 and 6.6. Samples were dissolved in phosphate buffer (pH 6.6), McIlvaine buffer (pH 4.6), and MRS media. MRS media was adjusted before sterilization at pH values of 4.6 and 6.6. Sterile centrifuge tubes were filled with 30 mL of sterile solution containing a final concentration of 200 $mgL^{-1}$ of the pyranoanthocyanin extract. Samples were kept at 35 °C for 7 d in darkness. Samples were taken at day 1, day 2, and at the end of the analysis. Quantification of analytes was done by UHPLC following the gradient shown in Table 4.18 in section 4.4.3.

### 4.3.10. Color measurement

To evaluate the total color difference of samples, color analyses were done. Color parameters were recorded as L*, a*, and b* CIELab coordinates values per triplicate. Calculation of chroma and hue angle from a* and b* values, are given by equation 1 and 2, respectively. In this system, L* coordinate correspond to lightness on a 0 to 100 scale going from black to white. While a* correspond to the degree of red (+) or green (-), and b* yellow (+) or blue (-) color. Chroma (C*) is used to evaluate the total color attributes. While, hue (h °) represent the color presented by the sample.

$$C^* = [(a*)^2 + (b*)^2]^{1/2} \quad \textbf{(1)}$$

$$h° = arctan\frac{b*}{a*} \quad \textbf{(2)}$$

For color evaluation, fermentation samples were first centrifuged at 8 000 rpm for 5 min to remove microorganisms, afterwards cuvettes were filled up with 15 mL of sample. Before the measurement, the colorimeter sensor was calibrated. The cuvette was placed directly onto the colorimeter sensor and each sample was measured in triplicate. Representation of total color differences ($\Delta E^*$) was done by Origin Pro 8G program (equation 3).

$$\Delta E^* = [(\Delta L)^2 + (\Delta a)^2 + (\Delta b)^2]^{1/2} \quad \textbf{(3)}$$

### 4.3.11. Lactic acid fermentation of BCJ

MRS media (pH 6.6) was used for cultivation of *L. plantarum*, *L. mali*, *L. sakei*, *L. helveticus* and *L. reuteri*, at 35 °C under shaking at 150 rpm, while, *L. fermentum* was grown at 37 °C. After 12 h cultures were used as inoculum. Prior to inoculation, bacteria were centrifuged for 5 min at 8 000 rpm and washed twice with sterile saline solution (0.9 % NaCl).

A multi-step reaction was used for synthesis of pyranoanthocyanins. Black carrot juices were fermented with binary strain combinations of *L. reuteri*, *L. helveticus* or *L. fermentum*, all of them possessing cinnamoyl esterase activity, with *L. plantarum*, *L. sakei* or *L. mali* showing hydroxycinnamic decarboxylase activity, respectively. Lactic acid bacteria, exhibiting esterase activity, were inoculated at the beginning of the fermentation, whereas lactic acid bacteria, showing decarboxylase activity, were added after 72 h of fermentation.

The inoculum was prepared by growing the strains overnight in 10 mL of MRS media at pH value of 6.6 and a temperature of 35 °C. Cells were washed twice with a sterile saline solution (0.9 % NaCl), centrifuged and re-suspended in 1 mL of sterile water for further use as inoculum. Juices were incubated and kept at 35 °C for 31 d, samples consisting of

1 mL were taken at the end of the fermentation for quantification of anthocyanins and pyranoanthocyanins by UHPLC-DAD (section 4.4.5).

### 4.3.12. Decarboxylation activities of *L. plantarum*, *L. mali*, and *L. sakei* in BCJ

Juice form BCJ variety deep purple was adjusted to pH 4.6 and 5.6 with lactic acid and acetic acid (4:1, v:v), respectively supplemented with CA at a concentration of 202 mM and used for fermentation. Black carrot juices were fermented for 48 h at 35 °C with 10 % of overnight cultures of *L. plantarum*, *L. mali*, and *L. sakei* grown in MRS media. BCJ acidified without inoculum was incubated at the same conditions and used as control. Samples were taken after 24 h and 48 h for quantification of fermentation products (section 4.4.5).

### 4.3.13. Evaluation of cinnamoyl esterase activities combined with phenolic decarboxylase activities of Lactobacilli

*L. mali* and *L. reuteri* or *L. plantarum* and *L. reuteri* were used as binary strain combinations for fermentation of BCJ. An inoculum of 10 % of overnight cultures of *L. reuteri* was taken, centrifuged at 8 000 rpm for 5 min and re-suspended in 5 mL of BCJ for esterase activities. After 24 h, samples were centrifuged, and the supernatants were recovered. BCJ was mixed with 5 mL of fresh juice and the pH was adjusted to a pH of 5.0 with NaOH 0.1 M, before being inoculated with 30 % of overnight cultures of *L. plantarum* or *L. mali*. Fermentations were performed at 37 °C under agitation. BCJ acidified at pH 4.0 without inoculum was incubated at the same conditions and used as control. After 24 h, samples were taken, centrifuged at 10 000 rpm for 5 min and the supernatant was membrane filtered for analysis (section 4.4.5).

### 4.3.14. Co-fermentation of BCJ to produce pyranoanthocyanins

2.5 mL of BCJ was inoculated with an inoculum consisting of 10 % of *L. reuteri* (J1). After 24 h J1 was centrifuged and the supernatant was recovered, 1.5 mL of fresh BCJ was inoculated with 30 % of *L. plantarum*, *L. mali* or *L. sakei*. The pellet of *L. reuteri* was re-inoculated in fresh 2.5 mL BCJ (J2). After 24 h J2 was centrifuged and the supernatant was mixed with J1, 3.5 mL of fresh juice were added to get a final volume of 10 mL. BCJ acidified with lactic acid and acetic acid (4:1, v:v) at pH 4.0 was used as a control. All fermentations were incubated at 37 °C for 48 h under shaking at 150 rpm. After 96 h, samples were taken for quantification of pyranoanthocyanins by UHPLC-ESI-MS$^n$ (section 4.4.5) Figure 4.2 shows the consecutive steps of this sequential fermentation.

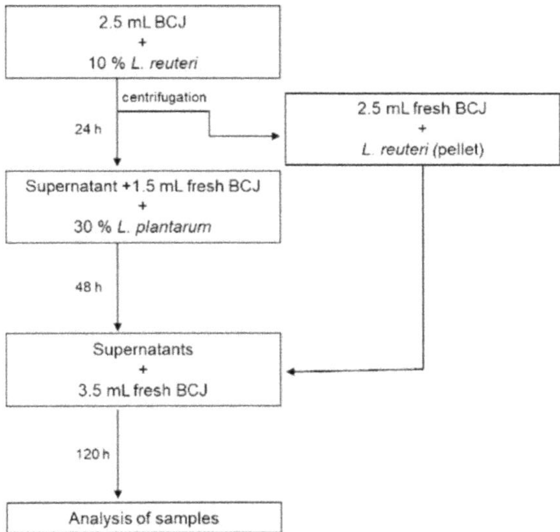


**Figure 4.2.** Co-fermentation of black carrot juice for the synthesis of pyranoanthocyanins by *L. reuteri* and *L. plantarum* at 37 °C

## 4.4. Analytical methods

### 4.4.1. Determination of the concentration

For all analysis carried out in this work, determination of the concentration of products was performed as shown in equation 4. Concentration can be defined as the mmol of fermentation products (FP) produced at a defined time ($t_x$) divided by the initial amount of hydroxycinnamic acids (HCA) and is expressed as a percentage.

$$\text{Concentration}(\%) = \frac{[\text{FP } t_x]}{[\text{HCA } t_0]} * 100 \qquad \textbf{(4)}$$

### 4.4.2. Determination of products from Lactobacilli activities at fermentation condition

Quantification of products from decarboxylation activities, esterase activities and tyrosine ammonia lyase activities by lactobacillus strains were performed by UHPLC-DAD. For analysis, the temperature of the column was maintained at 40 °C, the flow rate was set at 0.4 mL $min^{-1}$, and the injection volume was 2.5 µL. DAD detector was set at 280 nm for quantification of the corresponding vinyl derivatives and 320 nm for HCA. For separation, 0.1 % formic acid in water (v $v^{-1}$) was used as solvent A and 0.1 % formic acid in acetonitrile (v $v^{-1}$) as solvent B. Tables 15 and 16 show the gradients used for determination of the corresponding activities displayed by LAB. Quantification of *p*-CA, FA, CA, 5-CQA, CNA, L-Tyr, L-Phe, 4-VP, 4-VG and 4-EP were performed using standard calibration curves in a range of 0.1-1000 mg $L^{-1}$ with a $r^2 = 0.99$.

**Table 4.15.** UHPLC-DAD gradient for quantification of HCA and vinyl derivatives

| Time (min) | Eluent B (%) |
|---|---|
| 1.5 | 20 |
| 4.5 | 20 |
| 7.5 | 90 |
| 10 | 90 |
| 11 | 20 |

**Table 4.16.** UHPLC-DAD gradient for quantification of phenylpropanoid derivatives

| Time (min) | Eluent B (%) |
|---|---|
| 0 | 0 |
| 2 | 0 |
| 5 | 20 |
| 10 | 20 |
| 13 | 90 |
| 16 | 90 |
| 17 | 100 |
| 17.8 | 0 |
| 21 | 0 |

### 4.4.3. Determination of anthocyanins and anthocyanins adducts

For quantification of anthocyanins and pyranoanthocyanins, solvents composition was 5 % (v $v^{-1}$) formic acid in water and 5 % (v $v^{-1}$) formic acid in acetonitrile as eluent A and B, respectively. For analysis, the temperature of the column was maintained at 40 °C, the flow rate was set at 0.4 mL $min^{-1}$ and the injection volume was 5 µL. DAD detector was set at 520 nm for quantification of anthocyanins and pyranoanthocyanins. Quantification of products was performed using calibration curves of standards in a range of 0.1-1000 $mgL^{-1}$ with a $r^2 = 0.99$. Tables 4.17 and 4.18 show the gradients for separation of anthocyanins and pyranoanthocyanins.

**Table 4.17.** UHPLC-DAD gradient used for analysis of anthocyanins (method 1)

| Time (min) | Eluent B (%) |
|---|---|
| 0 | 4 |
| 2 | 4 |
| 7 | 8 |
| 13 | 10 |
| 19 | 17 |
| 23 | 30 |
| 23.3 | 100 |
| 25.3 | 100 |
| 25.8 | 4 |
| 30 | 4 |

**Table 4.18.** UHPLC-DAD gradient used for analysis of anthocyanins and pyranoanthocyanins (method 2)

| Time (min) | Eluent B (%) |
|---|---|
| 0 | 5 |
| 5 | 10 |
| 17 | 30 |
| 18 | 100 |
| 21 | 100 |
| 22 | 5 |
| 25.01 | 5 |

### 4.4.4. Determination of specific cinnamoyl esterase activity by UHPLC-ESI-MS$^n$

Quantification of fermentation products was carried out by UHPLC-ESI-MS$^n$. For analysis, solvents A and B consisted of water and acetonitrile, both acidified with 0.1 % (v $v^{-1}$) of formic acid. Column temperature was set at 40 °C, the flow rate used was of 0.4 mL $min^{-1}$ and 5 µL of sample solution was injected. A specific selected ion recording method (SIR) in negative ionization mode was used for ions at *m/z* 179, 353 and 515, which were attributed to CA, mono- and di-CQAs respectively, and followed by product ion scan yielding the respective fragments (*m/z* 135 for

CA, *m/z* 191 for CQA, and *m/z 353* for di-CQAs). Collision energy of 25 and 9 arbitrary units was set for CA and CQA, respectively. Quantification of CA, mono CQA, and di-CQAs was performed using calibration curves of 5-CQA and CA in a range of 0.1-500 $mgL^{-1}$ with a $r^2 = 0.99$, for quantification of di-CQAs, a molecular mass correction factor was applied. The following gradient (Table 4.19) was used for separation and quantification.

**Table 4.19.** Gradient used for quantification of CQA and CA

| Time (min) | Eluent B (%) |
|---|---|
| 0 | 2 |
| 1 | 2 |
| 3 | 3 |
| 4 | 8 |
| 12 | 12 |
| 20 | 95 |
| 22 | 95 |
| 23 | 2 |
| 25 | 2 |

### 4.4.5. Determination of pyranoanthocyanins generated during fermentation in BCJ by UHPLC-ESI-$MS^n$

For analysis, solvents A and B consisted of water and acetonitrile, both acidified with 0.1 % ($v\ v^{-1}$) of formic acid. Column temperature was set at 40 °C, the flow rate used was of 0.4 mL $min^{-1}$ and 5 µL of sample solution was injected. Anthocyanins in BCJ were identified and analyzed in positive ionization mode. For analysis, a collision energy of 35 arbitrary units was used, the capillary voltage was 4 V and the temperature equaled 325°C. Quantification of 5-CQA, CA, 4-VC, and Cy-3-glu products was performed using calibration curves of CA, 5-CQA, 4-VP, and Cy-3-glu in a range of 0.1-500 mg $L^{-1}$ with a $r^2 = 0.99$. For quantification of 4-VP and pyranoanthocyanins, a molecular mass correction factor was applied. Table 4.20 shows the gradient used for separation and quantification.

**Table 4.20.** Gradient used for quantification of fermentation products from BCJ

| Time (min) | Eluent B (%) |
|---|---|
| 0 | 10 |
| 15 | 25 |
| 18 | 90 |
| 21 | 100 |
| 24 | 100 |
| 24.01 | 10 |
| 28 | 10 |

### 4.5. Statistical analysis

To determine differences in substrate hydrolysis and product formation, an ANOVA followed by Tukey post hoc test with a level of significance of 95 % and a probability of $p \leq 0.05$ was performed using XLSTAT (Version 2014.4.06, AddinSoft Technologies, Paris, France). Results are expressed as the mean ± standard deviation.

# 5. RESULTS AND DISCUSION

## 5.1. Factors influencing cinnamate decarboxylase activity in lactobacilli

### 5.1.1. Effect of the pH value on the metabolism of free HCA

In food plants, HCA such as *p*-CA and FA are widely distributed (Herrmann & Nagel. 1989; Teixeira et al., 2013; Kabera et al., 2014) and several microorganisms are capable of decarboxylating HCA by the action of CD enzymes (Chatonnet et al., 1995; Couto et al., 2006; Curiel et al., 2010). Some LAB like *L. plantarum*, *L. mali*, and *L. sakei* are reported to possess high decarboxylation activity (Couto et al., 2006). The metabolic conversion of HCA by LAB involves two enzymatic activities. In the first stage, HCA is decarboxylated to their respective vinyl derivatives by CD enzymes, which can be further reduced by the action of VPR enzyme into ethyl phenols (Chatonnet et al., 1992; Chatonnet et al., 1995; Chatonnet et al., 1997; Couto et al., 2006). These metabolic activities in LAB are strongly affected by factors such as pH and, therefore, the objective of this study was the evaluation of CD activity in media supplemented with *p*-CA, or FA at four different pH-values (6.6, 5.6, 4.6, and 3.6) by *L. plantarum*, *L. mali*, and *L. sakei*.

All three LAB carried out decarboxylation of the free HCA (Figure 5.1). Microbial metabolism of *p*-CA and FA resulted in the formation of 4-VP and 4-VG, respectively. Production of vinyl derivatives is a detoxification mechanism of LAB that is stress-induced through HCA (Bel-Rhlid et al., 2013). In this study, the FA conversion rate of all LAB was found to be lower than *p*-CA conversion, independently of the pH-value (Figure 5.1 a and b). For both acids, higher decarboxylation rates yielding vinyl derivatives were obtained at pH values of 5.6 and 4.6 in comparison to those at pH 6.6.

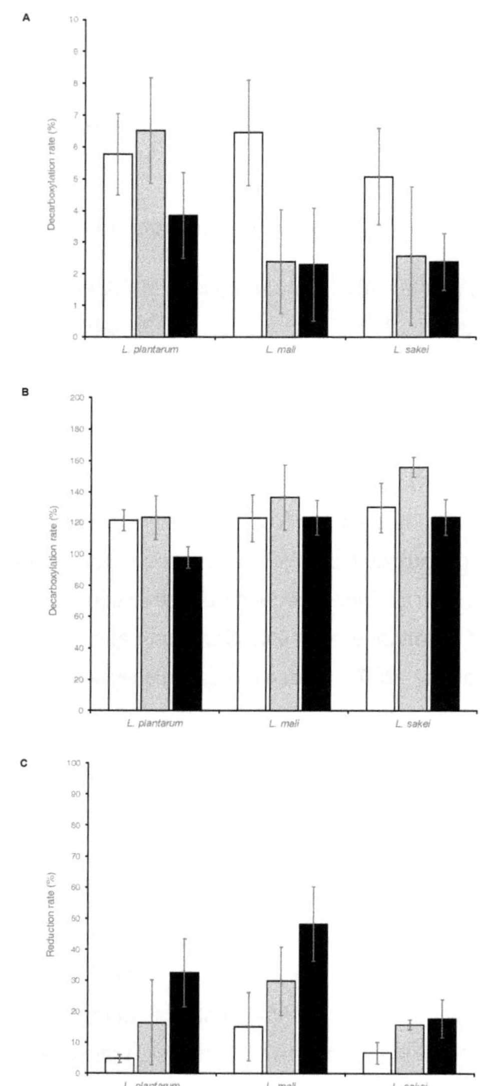


**Figure 5.1.** Influence of pH on CD activity to produce 4-VG from FA (A), 4-VP from *p*-CA (B) and 4-EP from 4-VP (C) at pH 4.6 (white bars), 5.6 (gray bars), and 6.6 (black bars)

Among the three LAB, the highest conversion rate of FA to 4-VG was obtained by *L. mali* (6.5 %) at pH 4.6. In the case of fermentation with *p*-CA, it was noticed that the conversion rate of *p*-CA to 4-VP was always over 100 % independently of the pH assessed (4.6, 5.6, and 6.6). This excess of 4-VP was calculated on basis of the initial amount of *p*-CA added for fermentation. Regarding these results, further experiments were carried out and confirmed that the excess could be due to an overestimation resulting from the standard used for quantification. Despite this overestimation, it was proven that *p*-CA showed a high preference for decarboxylation activity by LAB and was consumed in its totality. Overestimation of 4-VP was only present in experiments needing quantification and were always reproducible. All experiments to investigate the reason for this overestimation are explained in detail in section 5.2.

After fermentation, 4-EP was found in the media (Figure 5.1 c) formed after reduction of 4-VP by VPR enzymes. The highest production of 4-EP was obtained at pH 6.6. The results showed that 4-EP concentration increased when the pH-value increased. Even though *L. plantarum*, *L. mali*, and *L. sakei* can form low quantities of vinyl derivatives, fermentation with FA resulted in the formation of 4-VG but no 4-EG. This formation requires a reductase step, which has been identified only in a number of microorganisms, like *L. brevis*, *L. collinoides*, and *L. plantarum* (Cavin et al., 1993; Chatonnet et al., 1993; Couto et al., 2006).

In this work, cellular growth was used as a parameter for the evaluation of the inhibitory effect caused by HCA and the different pH-values. At concentration of 500 $mgL^{-1}$ of *p*-CA or FA, all pH-values tested act inhibitory on the growth of *L. plantarum*, *L. mali*, and *L. sakei*. The strongest inhibitory effect was at pH 3.6 where no growth was observed. Additionally, it was found that cellular growth at pH 6.6 was not affected compared to the control. Campos et al. (2009) reported an inhibitory effect of *p*-CA, FA, and CA on *L. hilgardii* growth at concentration of 500 $mgL^{-1}$ at pH 4.5. In contrast to the findings of this work, these authors reported *p*-CA to cause the highest inhibitory effects. Similar results were reported by Silva et al. (2011), who assessed *L. plantarum*, *L. collinoides,* and *Pediococcus pentosaceus* for the production of volatile phenols from *p*-CA at pH range of 3.5, 4.0, and 4.5. These authors reported higher production rates of 4-VP at pH 4.5. Meanwhile, at pH 3.5 only traces of 4-VP and 4-EP were found. Zaldivar et al. (1999) evaluated the effect of organic acids on the bacterial growth and found out that the toxicity of HCA was reduced when pH-value of the media increased. This was confirmed by Silva et al. (2011), who correlated these findings to the exposure to stress due to the pH-value and the presence of *p*-CA or FA, affecting the growth of LAB. The results of this study showed that decarboxylation rates decreased with increasing pH-value and vice versa. This happens because phenolic acids are found in their deprotonated form at low pH, facilitating their diffusion into the cell (Reguant et al., 2000). Until now, metabolization of phenolic acids has been suggested as a stress-induced response of microorganisms converting HCA into less toxic compounds (Spano & Massa, 2006; Sánchez-Maldonado et al., 2011; Bel-Rhlid et al., 2013). However, detailed information about this detoxification process of phenolic acids by microorganisms is still missing.

In LAB, the cell membrane works as a diffusion barrier between the cytoplasm and the media. This membrane is mainly responsible for maintaining functions essential for the survival of bacteria (McDonnell & Russell., 1999; Konings, 2002; Campos et al., 2009). Cell membrane in microorganism prevents external compounds from diffusing freely into the cell and prevent metabolites from being excreted in an uncontrolled way out of the cell, and, maintaining the electrochemical ion gradient is inherent to this function. As reported by Konings (2002), the respiratory chain protons are pumped out of the cell resulting in a pH gradient ($\Delta$pH), leading to an alkaline condition inside of the cell and acid outside. This generates an electric potential ($\Delta\psi$) product of the translocation of the positive charge from inside to outside. Those two gradients are the proton motive force. This proton motive force drives the energy necessary for cell processes, growth, and viability. Intracellular pH maintenance (approximately 7.5) is important to ensure cell viability. (Koning, 2002). Therefore, it is not surprising that optimal growth conditions for most LAB are found around neutral pH (6.0-6.5). A change in pH results in a decreased metabolic activity. Several metabolic processes could affect the internal pH, among them a decrease in external media pH due to fermentation activities. As a consequence, protons are pumped from the cytoplasm to the external media leading to an increase of internal pH (Konings, 2002; Campos et al., 2009). However, a major influx of protons results in more energy release, energy that is essential for other energy-requiring processes, directly impacting the cell viability. Nevertheless, at acidic fermentation conditions, this intracellular pH regulatory mechanism is not sufficient, and other proton release processes are required. In order to prevent the death of the cell through acid poisoning, proton fluxes catalyzed by proton-sodium or proton-potassium systems are activated. Intracellular acid accumulation can cause damage to the cell membrane, leading to a leakage of intracellular constituents to the external media. The release of potassium is reported as an indicator of cell membrane damage (McDonnell & Russell.,1999; Campos et al., 2009). In literature, pH-value is reported as factor inducing CD enzymatic activity (Campos et al., 2009; Silva et al., 2011). If bacterial growth is exposed to environmental stress, most of the survival functions rely on the cytoplasmic membrane. Therefore, it can be assumed that the excess of toxicity was responsible for the observed loss of cell viability at the lowest pH (pH 3.6). At low pH-values, HCA are mostly present in their dissociated form, which facilitates their diffusion into the cell. Diffusion of HCA may decrease the intracellular pH leading to cytoplasmic poison due to the acidification. The influx of protons out of the cell in order to restore the ΔpH causes a release of energy. Therefore, other metabolic functions can result which directly affect cell growth. This explains the observation of high decarboxylation rates at pH 5.6 and 4.6, which ensure the prevention of damage to the cytoplasm through poisoning. At pH 6.6, the low difference between intracellular pH and external pH decreased the toxic effect of HCA (figure

5.1). However, more research is necessary since full elucidation of this psychological mechanism has not yet been done.

### 5.1.2. Metabolism of *p*-CA, FA, and CA in MRS media and in buffer by LAB

For a better understanding of the CD activities of *L. plantarum*, *L. mali*, and *L. sakei* in the metabolism of phenolic acids, enzymatic activities were assessed in MRS media and phosphate buffer solution at pH value of 5.6 and 4.6. At these conditions, LAB showed high decarboxylation rates in the previous section (see section 5.1.1). Enzymatic activity was evaluated on basis of the amount of mM vinyl derivatives recovered from decarboxylation of HCA after 24 h of fermentation.

In this study, a formation of vinyl derivatives or ethyl derivatives was not detected in MRS media or buffer solution used as control. Fermentation products of HCA by LAB strains in MRS media or buffer solution are summarized in Table 5.1. The two-fold repetitions of the measurements showed only little deviation, as indicated by the standard deviation which never exceeded 12%. Despite the similar structure, HCA metabolism by LAB can be different (Campos et al., 2003; Campos et al., 2009; Sánchez-Maldonado et al., 2011). CD activity of all three LAB strains was higher to CA and *p*-CA than toward FA. In fermentations carried out in MRS media, all three LAB decarboxylated CA and *p*-CA completely into their respective vinyl derivatives, i.e., 4-VC, and 4-VP, respectively. As a result of missing substances necessary for bacteria survival, decarboxylation rates obtained in buffer solution were lower in comparison to MRS media. In MRS media, higher conversion rates of *p*-CA into 4-VP resulted at the lowest pH assessed. At pH 4.6, decarboxylation rates were found to be 163.79% for *L. plantarum* followed by *L. mali* and *L. sakei* with 154.42% and 152.85%, respectively. A maximum conversion rate of 2.77% of 4-EP was calculated for *L. sakei*. At an initial pH of 5.6, *L. mali* produced the highest amount of 4-VP (150.30%). In buffer solution, decarboxylation of *p*-CA was found to be lower than in MRS media.

**Table 5.1** Conversion rate and remaining of HCA by LAB in MRS medium and Buffer after 24 h of incubation at different pH-values

| | | pH 4.6 | | | pH 5.6 | | |
|---|---|---|---|---|---|---|---|
| **Lactobacilli** | **Media** | **Rem. *p*-CA** | **Increase VP** | **Increase EP** | **Rem. *p*-CA** | **Increase VP** | **Increase EP** |
| *L. plantarum* | Buffer | 76.17±4.08 $^{a}$ | 4.52±4.61 $^{f}$ | n.d. $^{d}$ | 68.14±2.64 $^{ab}$ | 21.92±7.58 $^{de}$ | n.d. $^{d}$ |
| | MRS | n.d. $^{c}$ | 163.79±0.14 $^{a}$ | 0.34±0.12 $^{cd}$ | n.d. $^{c}$ | 140.98±2.81 $^{c}$ | 0.93±0.18 $^{bcd}$ |
| *L. mali* | Buffer | 75.60±12.97 $^{a}$ | 4.36±0.70 $^{f}$ | n.d. $^{d}$ | 59.95±2.93 $^{b}$ | 30.22±0.51 $^{d}$ | n.d. $^{d}$ |
| | MRS | n.d. $^{c}$ | 154.42±1.55 $^{ab}$ | 1.85±0.86 $^{abc}$ | n.d. $^{c}$ | 150.30±3.72 $^{bc}$ | 2.53±0.52 $^{a}$ |
| *L. sakei* | Buffer | 76.37±1.59 $^{a}$ | 4.83±2.52 $^{b}$ | n.d. $^{d}$ | 70.08±0.27 $^{ab}$ | 19.21±4.74 $^{e}$ | n.d. $^{d}$ |
| | MRS | n.d. $^{f}$ | 152.85±1.70 $^{bc}$ | 2.77±1.17 $^{a}$ | n.d. $^{c}$ | 145.20±0.32 $^{bc}$ | 2.09±1.17 $^{ab}$ |
| | **Media** | **Rem. CA** | **Increase VC** | **Increase EC** | **Rem. CA** | **Increase VC** | **Increase EC** |
| *L. plantarum* | Buffer | 59.97±7.60 $^{bc}$ | 37.17±4.23 $^{cd}$ | n.d. | 72.34±4.71 $^{ab}$ | 26.88±2.96 $^{de}$ | n.d. |
| | MRS | n.d. $^{e}$ | 96.66±1.36 $^{a}$ | n.d. | n.d. $^{e}$ | 97.14±4.62 $^{a}$ | n.d. |
| *L. mali* | Buffer | 67.59±12.94 $^{abc}$ | 29.35±3.64 $^{de}$ | n.d. | 49.10±5.95 $^{cd}$ | 48.32±7.43 $^{bc}$ | n.d. |
| | MRS | n.d. $^{e}$ | 96.67±1.41 $^{a}$ | n.d. | n.d. $^{e}$ | 94.83±0.04 $^{a}$ | n.d. |
| *L. sakei* | Buffer | 81.85±2.58 $^{a}$ | 17.32±3.88 $^{e}$ | n.d. | 39.08±8.21 $^{d}$ | 58.45±7.59 $^{b}$ | n.d. |
| | MRS | n.d. $^{e}$ | 95.07±4.58 $^{a}$ | n.d. | n.d. $^{e}$ | 94.83±0.02 $^{a}$ | n.d. |
| | **Media** | **Rem. FA** | **Increase VG** | **Increase EG** | **Rem. FA** | **Increase VG** | **Increase EG** |
| *L. plantarum* | Buffer | 84.87±6.21 $^{a}$ | 7.05±0.55 $^{a}$ | n.d. | 87.44±8.74 $^{a}$ | 6.00±8.10 $^{a}$ | n.d. |
| | MRS | 86.43±3.10 $^{a}$ | 7.38±2.36 $^{a}$ | n.d. | 82.36±8.44 $^{a}$ | 6.23±4.24 $^{a}$ | n.d. |
| *L. mali* | Buffer | 85.39±4.72 $^{a}$ | 9.04±4.48 $^{a}$ | n.d. | 86.21±4.46 $^{a}$ | 5.91±0.76 $^{a}$ | n.d. |
| | MRS | 87.79±6.85 $^{a}$ | 8.73±1.87 $^{a}$ | n.d. | 89.67±2.88 $^{a}$ | 8.95±1.14 $^{a}$ | n.d. |
| *L. sakei* | Buffer | 81.61±9.31 $^{a}$ | 7.50±0.72 $^{a}$ | n.d. | 81.93±0.99 $^{a}$ | 6.69±4.01 $^{a}$ | n.d. |
| | MRS | 81.51±0.75 $^{a}$ | 7.21±4.23 $^{a}$ | n.d. | 82.75±1.48 $^{a}$ | 6.31±0.98 $^{a}$ | n.d. |

[a] Different letters represent significantly different results within the analyzed pH-value of the corresponding HCA ($p<0.05$); Values are mean ± standard deviation (n=2); n.d., not detected

The highest amount of 4-VP was obtained for *L. mali* at pH of 5.6 (30.22%) followed by *L. plantarum* (21.92%) and *L. sakei* (19.21%), respectively. All decarboxylation rates of CA were found similar for all three LAB strains and the highest was observed for *L. plantarum* (97.14%) at pH 5.6. The highest production rate of 4-VC in buffer solution was shown by *L. sakei* (58.45%) at pH 5.6, whereas at pH 4.6, *L. plantarum* had the highest yield (37.17%). Among all free HCA analyzed in this study, decarboxylation rate of FA was found to be the lowest for all strains. *L. mali* showed the highest yield of 4-VG in MRS media and buffer at pH of 4.6 with 8.73% and 9.04%, respectively. Although these LAB strains can further metabolize HCA until the formation of ethyl derivatives, 4-EP was only found in fermentations supplemented with *p*-CA in MRS media. However, the formation of 4-EC or 4-EG in the fermentations with CA and FA, respectively, was not detected, leading to the conclusion that the VPR activity was not present or active. Similar results were previously reported by Couto et al. (2006), who found no reduction activity of the 4-VG to 4-EG after investigating degradation of HCA by some LAB strains. Additionally, de las Rivas et al. (2009), who assessed CD activity by several LAB, ethyl derivatives were only found by fermentation conducted with *p*-CA while 4-VG from FA was not further reduced to 4-EG. The lower decarboxylation rate of FA compared to *p*-CA found in this work is similar to works previously published (Chatonnet et al.,1995; Van Beek & Priest, 2000). Cavin et al. (1993) evaluated the influence of FA and *p*-CA on the growth of some LAB. Although no quantification of products was shown, these authors confirmed the production of vinyl derivatives and ethyl derivatives after fermentation. The authors also reported a higher growth inhibition of LAB by FA compared to *p*-CA, observed by a longer lag phase. In comparison to other HCA, these authors associated the low metabolism of FA to a low tolerance and adaptive response reflected by the late induction of enzymatic activity. Couto el al. (2006) screened thirty-five strains of LAB for their ability to produce vinyl derivatives from *p*-CA and FA in culture media. The authors reported that *L. plantarum*, *L. mali*, and *L. sakei* converted *p*-CA to 4-VP in high quantities ranging from 80 to 110%. The conversion rate of 4-VP to 4-EP by these LAB was reported not to exceed 6%. The results obtained in this work show that stress-response to HCA was different between the tested LAB strains. The present results showing a preference towards *p*-CA and CA than to FA. suggest strain-specific differences as well as enzyme-substrate preference (Filannino et al., 2018). Such differences are responsible for the absence of VPR activity for generation of ethyl derivatives such as 4-EG and 4-EC. While the higher decarboxylation rates found at the lowest pH-value tested might be attributed to the stress generated by the pH media and presence of HCA (see section 5.1.1). Also, higher decarboxylation rates in media are probably higher due to the media composition which was rich in nutrients promoting bacterial growth.

### 5.1.3. Effect of phenolic acid concentration on Lactobacilli metabolism

In this study, the CD activities of *L. plantarum*, *L. mali*, and *L. sakei* toward different *p*-CA concentrations were studied. Fermentations were conducted in MRS media supplemented with concentrations between 0.1 mM and 5 mM.

Figure 5.3 shows the final product yields for different *p*-CA concentrations. After fermentation, no *p*-CA remained in the samples and products were detected in the control samples. Results showed that all three LAB strains convert *p*-CA into PhA at low concentrations. Metabolism of HCA like *p*-CA, CA, and FA has been shown to follow two metabolic pathways. They can be reduced to PhA, dihydrocaffeic, or dihydroferulic acid, respectively, or, CD activity can lead to the formation of vinyl derivatives (Barthelmebs et al., 2000; Curiel et al., 2010; Buron et al., 2011). As demonstrated in figure 2.9, one way involves the decarboxylation and further reduction of *p*- CA into 4-VP and 4-EP by CD and VPR enzymes. The second route implies PAR activity to form PhA which can be finally reduced to 4-EP by HPD enzymes (Cavin et al.,1997; Barthelmebs et al., 2000). Induction of both metabolic pathways is a consequence of the stress generated by the presence of phenolic acids, converting them into less toxic compounds. Furthermore, both pathways can co-exist and compete in the metabolism of HCA. The same products were identified by Cavin et al. (1997). These authors reported two inducible enzymes presented in *L. plantarum* which are involved in the metabolism of *p*-CA.

At low *p*-CA concentrations (0.1 mM, 0.2 mM, and 0.3 mM), 4-EP and PhA were quantified as main fermentation products of *L. plantarum* and *L. sakei*, whereas, no 4-VP was found in media. The 4-EP concentrations suggest that reductase activity of 4-VP towards formation of 4-EP is preferred at lower concentrations (< 1 mM). At these concentrations, *L. mali* was the only bacteria showing presence of 4-VP after fermentation. Fermentation products of *L. mali* were found to be different in comparison with products of *L. plantarum* and *L. sakei*. At concentration of 0.1 mM, *L. mali* produced PhA followed by 4-EP and 4-VP. At 0.2 mM, and 0.3 mM, the product concentration followed the order PhA > 4-VP > 4-EP. Additionally, *L. mali* showed the maximum formation of PhA among all three LAB used with 94.8% at a concentration of 0.1 mM.

At concentrations above 1 mM, formation of 4-VP was favored over 4-EP and PhA by *L. plantarum*, *L. mali*, and *L. sakei*. Therefore, it could be noticed that these LAB tend to favor decarboxylation into 4-VP over formation of PhA at high *p*-CA concentrations of *p*-CA (2 mM, 3 mM, and 5 mM). These results suggest that decarboxylation of *p*-CA into 4-VP occurs at higher concentrations (> 1 mM), whereas, the induction of a second metabolic route, producing PhA occurs at concentrations below 1 mM. At these concentrations all three Lactobacilli showed a very limited capacity to produce PhA. However, in fermentations by *L. sakei*, low amounts of PhA were quantified. These results suggest that at concentrations above 1 mM, CD activity towards 4-VP has priority over reduction into PhA. In contrast to this effect, at concentrations below 1 mM, PhA or 4-EP appeared to be favorably produced.

The results suggest that initial concentration plays an important role in the formation of fermentation products. Under the experimental condition of this study, all three LAB strains showed ability to induce PAR enzymes at low concentration of *p*-CA. PhA, and 4-VP were mainly found as fermentation products, while high concentration of *p*-CA favored the decarboxylation into 4-VP and 4-EP. Campos et al. (2003), reported HCA to have a growth inhibitory effect on LAB which might increase with concentration. Additionally, CD activity in presence of HCA is reported as a response mechanism to rest toxicity. In this respect, Barthelmebs et al. (2000) found PhA as a product of *p*-CA metabolism during the identification of an alternative CD pathway, present by knockout of the *p*-coumarate decarboxylase gene of *L. plantarum*.

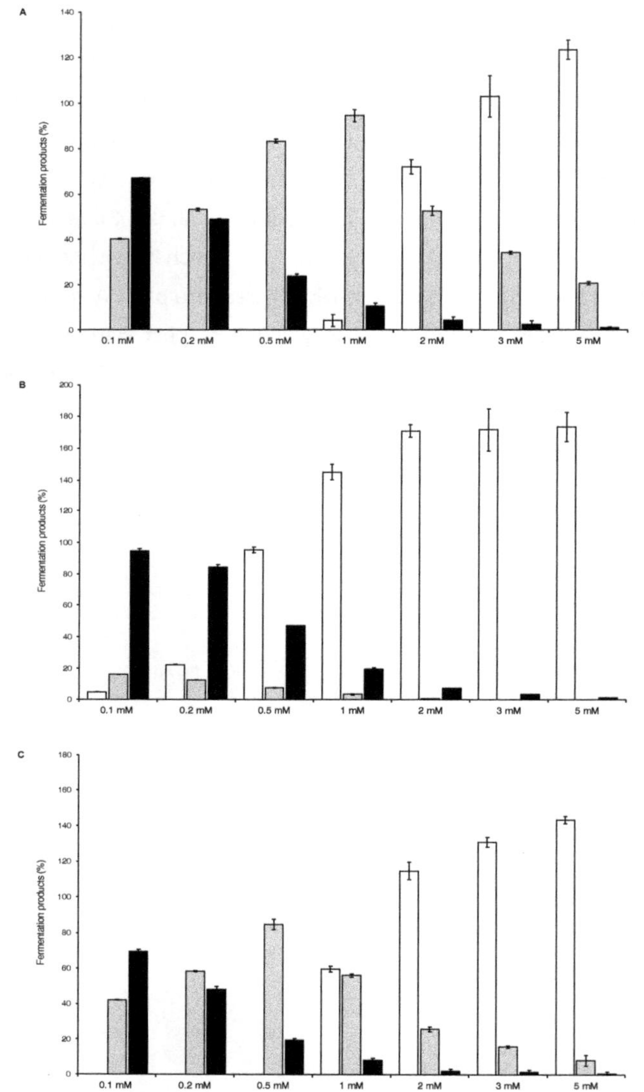


**Figure 5.3.** Quantification of product yields by *L. plantarum* (A), *L. mali* (B), and *L. sakei* (C) at different concentrations of *p*-CA. 4-VP shown in white bars, 4-EP in gray bars, and PhA in black bars

This alternative PAR enzyme was able to reduce phenolic acids into phenyl propionic acids, i.e., PhA, hydroferulic acid, or dihidroxycaffeic acid from *p*-CA, FA, and CA, respectively. These authors reported PAR activity responsible for the formation of PhA to be less efficient in decreasing toxicity of *p*-CA.

Thus, this metabolic pathway was probably followed by all three LAB at low concentration of *p*-CA. Whereas, higher concentrations of *p*-CA may lead to a decrease of external and intracellular pH, damage of cell membrane, and reduction of the proton motive force. As previously discussed, cellular membrane is responsible for bacteria survival and, at acidic conditions, an influx of protons out of the cell is activated in order to restore the ΔpH causing a release of energy. It is reported that transport processes are energetically expensive consuming metabolic energy needed for other processes (Konings, 2002). Additionally, Filanino et al. (2014), proposed LAB to use HCA as an external acceptor of electrons. These authors reported this reaction as strategy to avoid the hostile environment generated in the presence of HCA. This hypothesis proposed for the metabolism of HCA involves the reoxidation of the reduced cofactor NADH, providing energy through $NAD^+$ regeneration. Thus, LAB reduce 4-VP to 4-EP by increasing the availability of $NAD^+$. Therefore, the metabolic pathway leading to 4-VP and 4-EP production might be more efficient for reducing the toxicity of *p*-CA.

### 5.2. Potential presence of TAL or PAL activities

In previous experiments (see section 5.1), the calculated conversion rate of *p*-CA to 4-VP by CD activity exceeded 100%. This excess of vinyl derivatives was only found in fermentations supplemented with *p*-CA. Vinyl derivatives were quantified on the basis of the initial substrate concentration. Additionally, results at different concentrations of *p*-CA showed an increase on 4-VP formed by increasing the concentration of *p*-CA. This excess 4-VP was thought to be the result of microbial activity using L-Tyr or L-Phe as substrate by TAL or PAL enzymes.

Therefore, characterization of enzymatic activities involved in the formation and degradation of *p*-CA by LAB was investigated.

### 5.2.1. Enzymatic activities involve in the metabolism of *p*-CA

In plants PAL and TAL, enzymes are responsible for HCA synthesis. PAL enzyme is involved in the formation of trans-cinnamic acids, while TAL enzyme is highly selective for the synthesis of *p*-CA from the amino acid L-Tyr (Watts & Mijits., 2006; Louie & Bowman., 2006; Xue et al., 2007a; Kang et al., 2012; Jung et al., 2013). The microbial route to produce HCA from glucose can take place through the deamination of amino acids, such as L-Phe and L-Tyr by PAL or TAL enzymes, respectively (Watts & Mijits., 2006; Louie & Bowman., 2006; Xue et al., 2007a, 2007b; Kang et al., 2012). For this reason, L-Tyr, L-Phe, CNA, PhA, *p*-CA, 4-VP, and 4-EP were chosen for this study.

Results showed that all three LAB strains yielded 4-VP and 4-EP from fermentations with *p*-CA (Figure 5.4).

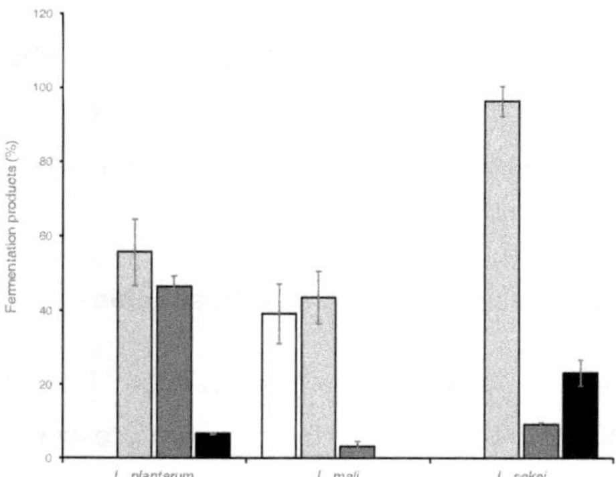


**Figure 5.4.** Fermentation products of *p*-CA by *L. plantarum*, *L. mali*, and *L. sakei*. Shown in white bars the remaining *p*-CA, 4-VP in slight gray bars, 4-EP in dark gray bars, and PhA in black bars

Additionally, PhA was found in fermentation samples of *L. plantarum* and *L. sakei.* These products have already been identified by several authors, since *p*-CA can be metabolized through two different pathways (Cavin et al.,1997; Barthelmebs et al., 2000; Filannino et al., 2014). Similar results were reported by Sánchez-Maldonado et al. (2011) who identified PhA, dihydrocaffeic acid, and dihydroferulic acid after assessing the metabolism of HCA by *L. plantarum*, and *L. fermentum* (1 mM for 24 h). In this study, PhA was not identified in fermentation by *L. mali.* A similar result was reported by Curiel et al. (2010), who only detected 4-VP from fermentation with *p*-CA by *L. brevis.* In fact, the metabolic pathway followed by LAB strains for detoxification of HCA like *p*-CA can mainly be explained by strain-specific differences (Filannino et al., 2018). A previous publication by Filannino et al. (2014) reported HCA to be used by LAB as external acceptor of electrons involved in the generation and preservation of the energy balance to avoid the toxic effect produced by HCA. Probably *p*-CA presented less toxicity to *L. plantarum* and *L. sakei* in comparison of *L. mali*, which metabolized *p*-CA through the decarboxylation pathway. This last route reported to be more efficient to detoxify this compound (Barthelmebs et al., 2000)

Experiments conducted with 4-VP showed a VPR activity yielding formation of 4-EP, but no further reaction was identified (Figure 5.5). The present findings confirm the results reported by previous authors, that the decarboxylation route followed by LAB form 4-EP as ending product from *p*-CA (Cavin at al., 1993; Chatonnet et al., 1997; Van Beek & Priest, 2000; Couto et al., 2006; de las Rivas et al., 2009; Curiel et al., 2010).

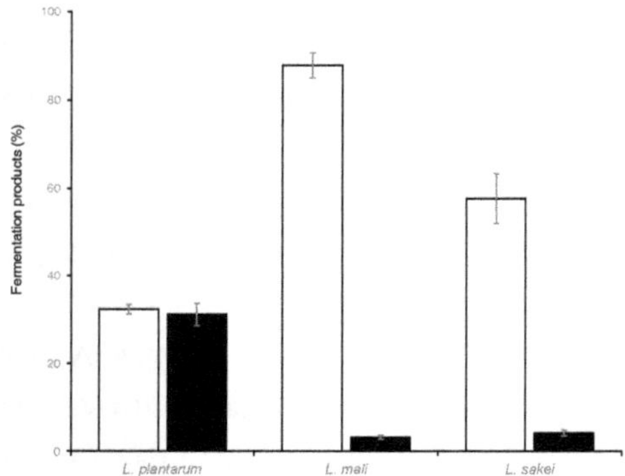


**Figure 5.5.** Quantification of enzymatic products by *L. plantarum*, *L. mali*, and *L. sakei* in fermentations with 4-VP. Remaining 4-VP after fermentation is shown in white bars, and 4-EP in the black bars

Figure 5.6 shows the remaining substrate concentration of 4-EP, PhA, and CNA in MRS after fermentation. Quantification showed a decrease in their concentrations. However, no metabolic products were identified for all three LAB when 4-EP, PhA, CNA, L-Tyr, or L-Phe was applied as substrate. An important decrease in CNA concentration in MRS media by all LAB was found after fermentation. A complete consumption was observed in experiments conducted with *L. sakei*. However, *p*-CA, 4-VP, 4-EP, or other enzymatic products were not detected. This rules out the possibility that the excess of 4-VP was produced after the decarboxylation of previously formed *p*-CA through the metabolism of CNA by all strains. Nevertheless, it is also possible that CNA might be decarboxylated into L-Tyr and further reduced to ethylbenzene by possible CD enzymes (Shimada et al., 1992; Middelhoven & Gelpke, 1995; Richard et al., 2015). Those activities have been identified in several microorganisms, Shimada et al. (1992) observed the formation of L-Tyr by *Pichia carsonii* in fish products using CNA as an antimicrobial agent. Middelhoven and Gelpke (1995) reported the conversion of CNA into L-Tyr by growing culture and cell-free extracts of *Cryptococcus elinovii*, whereas other studies, carried out with *Saccharomyces cerevisiae*, displayed CD activity, yielding L-Tyr and dihydroxycinnamic acid as metabolites of CNA (Larsson et al., 2001). However, information about the metabolism of CNA by Lactobacilli has not been reported to date. Detection of L-Tyr and L-Phe amino acids was not possible by the UHPLC method applied. However, separate analysis of L-Tyr and L-Phe was carried out by UHPLC-MS to find out if some products could be formed after fermentation. TAL activity would produce *p*-CA after deamination of L-Tyr, whereas PAL activity catalyzes the elimination of ammonia from L-Phe into CNA. Then, CNA might be converted into *p*-CA via a cinnamic acid 4-hydroxylase. However, in the samples of the fermentations with L-Tyr and L-Phe, no enzymatic products were found. Similar results were reported by Groot & de Bont (1998), these authors suggested the production of benzaldehyde by the action of PAL enzymes with the formation of CNA as intermediate. Nevertheless, no intermediates from L-Phe consumption were detected. The results obtained in this study with L-Tyr and L-Phe were not unexpected since these amino acids are the final products of the shikimate pathway present in bacteria (Rodriguez et al., 2014). Additionally, PAL or TAL activities have been mainly reported by engineered microorganisms (Qi et al., 2007; Vannelli et al., 2007; Xue et al., 2007b; Sariaslani, 2007; Gosset, 2009; Vargas-Tah et al., 2015). However, no PAL or TAL activities are known for Lactobacillus.

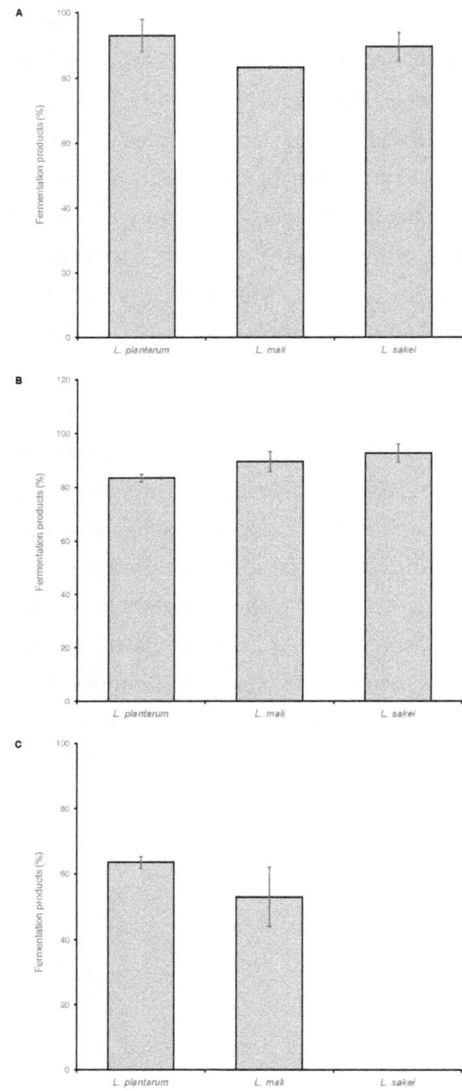


**Figure 5.6.** Remaining substrate concentration 4-EP (A), PhA (B), and CNA (C) after fermentation

### 5.2.2. Production of *p*-CA by TAL activities displayed by *S. espanaensis*

There are reports of a few microorganisms like *Rhodobacter sphaeroides*, *Rhodobacter capsulatus* or *Saccharothrix espanaensis* able to carry conversion to *p*-CA from L-Tyr as substrate (Singh et al., 2000; Louie et al., 2006; Xue et al., 2007b; Kang et al., 2012). For these reasons, *S. espanaensis* was used as positive control. After fermentation, a peak was detected by UHPLC-DAD and identified as *p*-CA.

### 5.2.3. PDC activities in PMM7 minimal media

In this study PMM7 minimal media was used to gain a deeper understanding of TAL, or PAL activities and CD activity of *L. plantarum*, *L. mali*, and *L. sakei*. This media was chosen because it provides the basic requirements for lactobacilli growth (Wegkamp et al. 2010).

LAB have high nutritional requirements for their growth, i.e., carbon and energy source, amino acids, vitamins, mineral, and nucleic acids (Konings, 2002). In previous experiments *L. plantarum*, *L. mali*, and *L. sakei* were grown in MRS media, which is a complex media recommended for the cultivation of Lactobacilli. Due to its undefined chemical composition, it is difficult to establish the effect on CD activity since some compounds, e.g., amino acids, can be involved in the synthesis of HCA. Additionally, through glucose conversion by TAL and/or PAL enzymes, formation of HCA, i.e., *p*-CA, and CNA, respectively, is promoted. Therefore, the effect of different culture media components needed to be investigated to determine their influence on TAL or PAL activity. PAL and TAL activities by *L. plantarum*, *L. mali*, and *L. sakei* were assessed in minimal media containing L-Tyr and L-Phe. An additional fermentation in minimal media without the both amino acids L-Tyr and L-Phe was performed with *L. plantarum*.

The results showed that *L. plantarum*, *L. mali*, and *L. sakei*, decarboxylate *p*-CA in PMM7 minimal media similar than in MRS media. In all three concentrations and for all three LAB strains, 4-VP and PhA were obtained as unique fermentation products (Figure 5.7). The results showed decarboxylation of *p*-CA yielding 4-VP as the principal route of detoxification, while PhA formation was greater at the lowest *p*-CA concentrations (0.2 mM). In this minimal media, no reductase activity was exhibited. This suggests that reductase activity is dependent on the media composition. *L. plantarum*, *L. mali*, and *L. sakei* showed similar behavior, an increment in *p*-CA concentration resulted in more 4-VP formed, while a decrease of PhA formation was observed. At concentration of 0.2 mM, the maximum 4-VP production with 106% was observed for *L. plantarum* (figure 5.7). At concentrations of 2mM, and 5 mM, 4-VP exceeds 100% for all three LAB strains.

Fermentations by *L. plantarum* in the absence of both amino acids resulted in the formation of 4-VP as the only product. Again, decarboxylation rates of *p*-CA were found to be greater than 100%. Thus, it can be concluded that these amino acids do not serve as a substrate for HCA formation, and, therefore, do not increase the substrate concentration (Figure 5.8). After 48 h, a maximum *p*-CA decarboxylation rate of 191 % was observed. These findings allow the discarding of MRS media being involved in the excessive 4-VP production described in previous sections (see section 5.1). Additionally, results of experiments carried out in PMM7 minimal media indicate non-presence of TAL/PAL activities for all three LAB strains.

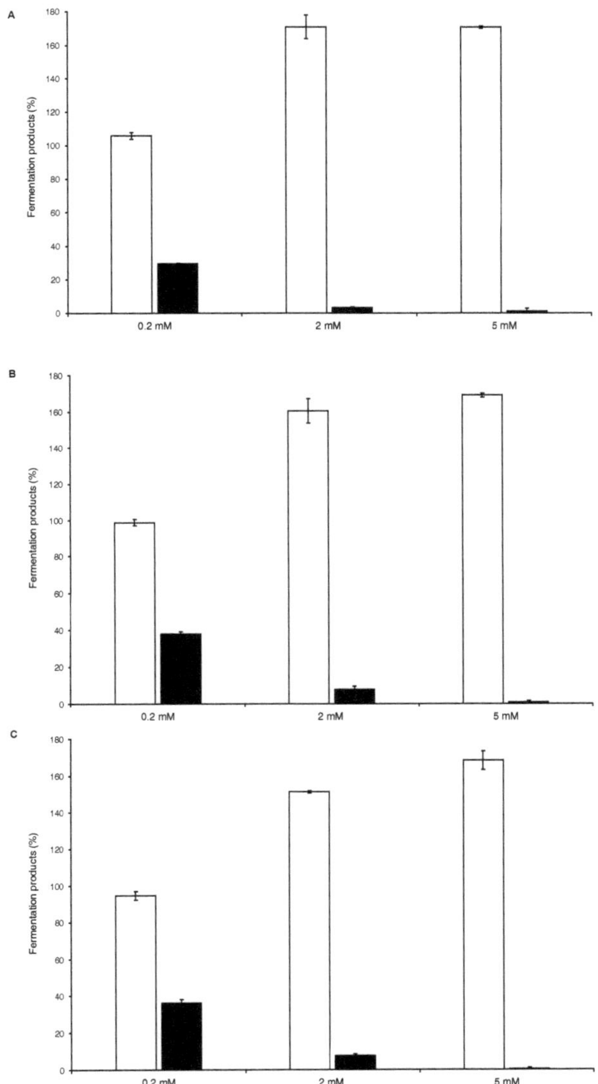


**Figure 5.7**. Metabolic products at different substrate concentrations of *p*-CA by *L. plantarum* (A), *L. mali* (B), and *L. sakei* (C) in PMM7 minimal media. 4-VP shown in white bars and PhA in black bars

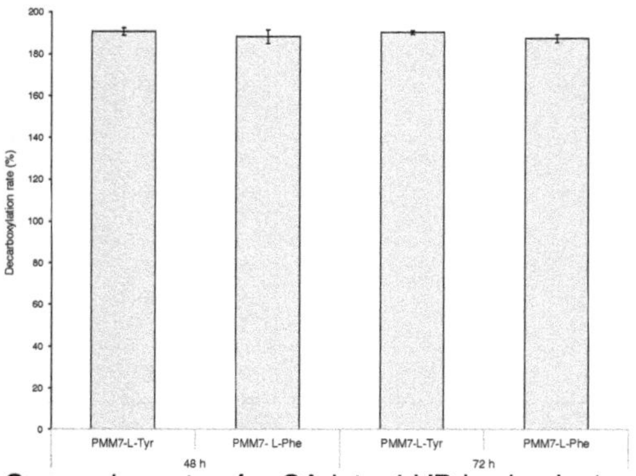


**Figure 5.8.** Conversion rate of *p*-CA into 4-VP by *L. plantarum* in PMM7 minimal media in absence of L-Tyr or L-Phe after 48 h and 72 h of fermentation

### 5.2.4. PDC activities in phosphate buffer

After the excess in conversion rate of 4-VP quantified in fermentation carried out in complex MRS media and MPP7 minimal media, metabolism of *p*-CA by *L. plantarum*, *L. mali*, and *L. sakei* was evaluated in phosphate buffer (pH 6.6) after the excessive production of 4-VP quantified in fermentations carried out in complex MRS media and MPP7 minimal media. Use of phosphate buffer allows the ruling out the use of specific substances, such as L-Tyr or L-Phe, both present in the media since they could be used for the synthesis of *p*-CA by LAB

After fermentation, no fermentation products were identified in the fermentation control. All three LAB yielded 4-VP as the only fermentation product. However, conversion rates exceeded the 100 %. Quantification of 4-VP after 48 h was 175.4%, 180.5%, and 154.5% for *L. plantarum*, *L. mali*, and *L. sakei*, respectively. Even with the high conversion rates, *p*-CA was found remaining in buffer after fermentation. As expected, no bacterial growth was found after fermentation since phosphate buffer does not contain substances necessary for bacteria growth.

Regarding the results in all previous experiments in section 5.2, it can be suggested that the excess of 4-VP quantified appears to be an analytical error. Considering the results obtained in section 5.1.3 at different *p*-CA concentrations, 4-EP and PhA were found as the only products in fermentations supplemented with *p*-CA at concentrations below 1 mM while 4-VP was quantified at concentrations above 1 mM. Additionally, fermentations, carried out with PhA (section 5.2.1) as sole substrate, did not show formation of any products. Based on these results, it is possible to confirm that below 1 mM, *p*-CA is partly reduced to PhA and decarboxylated into 4-VP which is further reduced to 4-EP. In Figure 5.3 in section 5.1.3, it can be seen that fermentation products formed below 1 mM of *p*-CA are about 100%. Excess in fermentation products was only obtained when 4-VP was found in media and its quantification was required.

Taking into consideration the evaluation of enzymatic activities involved in the metabolism of *p*-CA (section 5.2.1), the use of 4-VP as sole substrate confirmed the formation of 4-EP as the only product. Figure 5.5 shows *L. plantarum* with the highest reduction of 4-VP to 4-EP. The conversion rate of 4-VP to 4-EP was reported to be 32.4%, while 31.1% of 4-VP was remaining in the media after fermentation. This generates a difference of 36.5% of the initial concentration that is missing. It is known that 4-EP is the end product of PAD activities ruling out the consumption of this for other metabolic reactions. Therefore, these results may indicate an error in the estimation of 4-VP concentration with a possible overestimation. Overestimation of 4-VP after quantification of PAD activities by Lactobacilli was already reported (Soares, 2014). 4-VP is a highly reactive substance and available commercial reference compound has a concentration of 10% (w/w) according to the manufacturer. Soares (2014) reported an excess of 4-VP similar to the rates obtained in the present work using gas chromatography. Regarding this, quantification was done by other methods different to the one used in this work. Hence, although both works have in common the use of the commercial substance for quantification, a possible error in quantification due to the concentration found in commercially standard cannot be ruled out.

### 5.3. Determination of cinnamoyl esterase activity

HCA such as *p*-CA, FA, and CA are present in a large variety of food-plants (Couteau et al., 2001). However, in nature, HCA are mainly found as ester conjugates (Herrmann & Nagel. 1989; Crepin et al., 2004; Manach et al., 2004; Torres-Mancera et al., 2011; Teixeira et al., 2013; Bel-Rhlid et al., 2013). Release of free HCA from these esters is carried out by cinnamoyl esterase (CE) enzymes. This group of enzymes comprises carboxyl ester hydrolases able to hydrolyze the ester bond present in hydroxycinnamic esters releasing free HCA (Crepin et al., 2004; Topakas et al., 2007; Abeijón Mukdsi et al., 2012; Tomaro-Duchesneau et al., 2012). In humans and animals, HCA cannot be hydrolyzed by enzymes present during digestion reaching the colon almost intact. In the large intestine, they can be hydrolyzed by microbiota possessing CE enzymes (Hole et al., 2012). Among the wide range of microorganisms possessing CE enzymes, some bacteria, e.g., *L. fermentum*, *L. helveticus*, and *L. reuteri* have been also identified within the human microbiota (Topakas et al., 2007; Lai et al., 2009; Esteban-Torres et al., 2015). In this chapter, *Lactobacillus* strains were investigated regarding their CE activity and the substrate-specificity displayed by these enzymes.

#### 5.3.1. Determination of cinnamoyl esterase activity on 5-CQA by LAB

The most widespread HCA are the family of CQA (Gonthier et al., 2003; Bel-Rhlid et al., 2013), where 5-CQA is found as the most abundant compound (Clifford, 2000b; Deshpande et al., 2014;). Therefore, 5-CQA was used for the evaluation of CE activities displayed by some LAB strains. The determination of CE activity is defined as the amount of CA produced after cleavage of 5-CQA. In this study, the influence of pH on CE activity was evaluated for *L. fermentum*, *L. reuteri*, *L. johnsonii*, *L. helveticus*, *L. acidophilus*, and *L. gasseri*.

*L. johnsonii*, *L. acidophilus*, and *L. gasseri* did not show CE activity since these strains were not able to carry out the hydrolysis of 5-CQA. As reported by Filannino et al. (2018), the metabolism of phenolic acids is a strain-specific activity. Meanwhile, *L. helveticus*, *L. reuteri*, and *L. fermentum* strains were able to hydrolyze 5-CQA. CA was identified as product of CE activity since no formation of CA was observed in the control fermentations. Table 5.2 presents the hydrolysis rates of 5-CQA into CA. At both pH assessed (6.6 and 5.6), hydrolysis rates higher than 50% were obtained. However, all LAB strains showed higher CE activities at the higher pH-value tested (6.6). *L. reuteri* displayed the highest production of CA at both pH-values with 85.29 % and 76.04 %, respectively, for 6.6 and 5.6. Similar results are found in literature by authors who analyzed the influence of pH and temperature in CE activity (Topakas et al., 2007; Brod et al., 2010; Bel-Rhlid et al., 2013; Fritsch et., 2017).

The temperature used in this study (35 °C) was chosen to facilitate CE activities and bacterial growth. In literature, the optimal temperature for CE activity has been reported between 30 °C to 65 °C and a pH range from 4 to 8 (Topakas et al., 2007; Brod et al., 2010; Fritsch et., 2017). Fenster et al. (2003) characterized an esterase gene isolated from *L. casei* LILA, which showed the highest activity at 30 °C and pH between 5.5 to 6.0. Studies regarding the evaluation of the effect of pH and temperature on CE activity by *L. plantarum* carried out by Brod et al. (2010) resulted in high stability at temperatures ranging from 30 to 50 °C. Nevertheless, Szwajgier et al. (2011) reported the highest CE activity at 37 °C. Additionally, Esteban-Torres et al. (2013) analyzed CE activity showed by *L. plantarum* in a pH range from 3 to 9 at 20 °C. These authors studied CE activity in pH ranging from 3 to 8 reporting the highest activity at pH 6.5. Fritsch et al. (2017) did a characterization of CE activities from different lactobacillus and bifidobacterial strains.

**Table 5.2.** Hydrolysis of 5-CQA to CA by *L. fermentum*, *L. reuteri*, and *L. helveticus* after 24 h of fermentation [a,b,c]

| | pH 5.6 | | pH 6.6 | |
|---|---|---|---|---|
| **Lactobacilli** | **Remaining CQA %** | **Increase CA %** | **Remaining CQA %** | **Increase CA %** |
| *L. fermentum* | 43.80 ± 1.76 [a,1] | 52.65 ± 3.28 [b,2] | 29.40 ± 2.93 [c,1] | 64.03 ± 0.63 [c,2] |
| *L. helveticus* | 37.52 ± 1.21 [b,2] | 53.53 ± 4.78 [a,2] | 14.07 ± 0.09 [d,2] | 80.54 ± 1.46 [b,1] |
| *L. reuteri* | 17.54 ± 0.38 [d,3] | 76.04 ± 6.65 [a,1] | 14.08 ± 0.23 [d,2] | 85.29 ± 1.62 [a,1] |

[a] Values are mean ± standard deviation (n = 3)
[b] Means in the same line with a different letter are significantly different ($p < 0.05$)
[c] Means in the same column with a different number are significantly different ($p < 0.05$)

These authors reported that up to 35 °C, *L. plantarum* showed a high CE activity. Regarding the effect of the pH-value on CE activities by *L plantarum*, *L. reuteri*, *L. gasseri*, *L. acidophilus*, *L. helveticus*, and *L. fermentum* these authors analyzed CE activities in a pH range from 4 to 9. The results indicated a high CE activity at a range of 7.0 to 8.0, except for *L. helveticus*, which exhibited the highest activity at 8.5. Whereas Bel-Rhlid et al. (2013) evaluated the effect of the pH and temperature on the formation of CA and 4-VC by *L. johnsonii*. They reported no significant difference at 30 °C or 37 °C. These authors also reported no significant difference for the formation of CA between pH 5.0 and 6.0. In this study, they also found that *L. johnsonii* displayed CE and CD activities at 37 °C with the formation of CA and 4-VC after hydrolysis of CQA. However, in the present work, *L. fermentum*, *L. helveticus*, and *L. reuteri* (CD, PAR or VPR activities) conducted the formation of 4-VC, dihydrocaffeic acid, or 4-EC after hydrolysis of 5-CQA. This high CE activity displayed by *L. fermentum*, *L. helveticus*, and *L. reuteri* is of interest while the released products are of potential industrial interest. Free HCA have a wide variety of applications such as in the biofuel, cosmetic, pharmaceutical, or food industries (Dilokpimol et al., 2016)

### 5.3.2. Specific cinnamoyl esterase activity from monochlorogenic acids and dichlorogenic acids

In this study, *L. helveticus*, *L. reuteri*, and *L. fermentum* were grown in MRS media containing different monochlorogenic or dichlorogenic acids and tested for their cinnamoyl esterase specificity. Monochlorogenic fraction 3-CQA, 4-CQA, and 5-CQA were isolated from green coffee beans, while Jerusalem artichoke was used a source of di-CQAs. Table 5.3 shows the identification of the different fractions. Previous works have identified CQA presented in green coffee beans and Jerusalem artichoke (Schütz et al., 2004; Clifford et al., 2006; Kapusta et al., 2013; Deshpande et al., 2014; Chen at al., 2014). In addition, variation of the di-CQAs presented in Jerusalem artichoke reported in this study may present some differences compared to the results given in the literature due to isomerization of CQA during extraction. Monochlorogenic acid fractions presented the corresponding *m/z* 353 and the fragments ions at *m/z* 179 and *m/z* 191 corresponding to the CA and quinic acid moieties respectively. Dicaffeoylquinic acids, showed their precursor ion at *m/z* 515 rendering a fragment ion at *m/z* 353. Three monochlorogenic acids were identified in green coffee bean extracts. Isolation was performed by HPCCC. Figure 5.9 shows the fractions recovered after isolation by HPCCC, Fraction 1 consisted of a mixture of chlorogenic acids, fraction 2 contained only 3-CQA, fraction 3 was a mixture of 3-CQA, 4-CQA, 5-CQA, and 3-FQA; in fraction 4, 4-CQA and 5-CQA were detected, and fractions 5 and 6 contained pure 3-FQA and 5-CQA, respectively. After separation, di-CQAs remained in the coil. The di-CQAs that remained after separation with Jerusalem artichoke were used for fermentation. The culture conditions, temperature, and CQA concentration used in this study were chosen to facilitate bacterial growth and induction of enzymes.

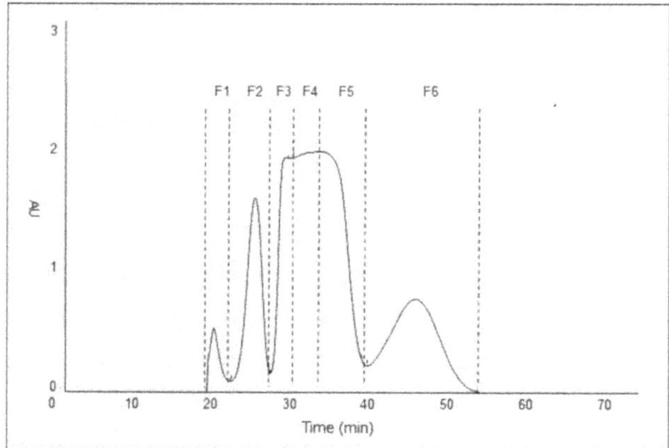


**Figure 5.9.** HPCCC chromatogram of the extract of green coffee beans: (F1) mixture of CQAs, (F2) 3-CQA, (F3) 3-CQA, 5-CQA, 3-FQA and 4-CQA, (F4) 4-CQA and 5-CQA, (F5) 4-FQA, (F6) 5-CQA. Chromatogram recorded at 324 nm

**Table 5.3.** Fragmentation data for monochlorogenic acids and dichlorogenic acids [a]

| CQA | Product ion *m/z* | $MS^2$ base peak *m/z* (abundance) | $MS^3$ base peak *m/z* (abundance) |
|---|---|---|---|
| 3-*O*-caffeoylquinic acid | 353.1 | 191.1(100), 179.1 (49), 135.1 (10) | 353.1→85.1, 173.1 (95), 127.1 (84), 93.1 (48), 111.2 (22) |
| 4-*O*-caffeoylquinic acid | 353.1 | 173.0 (100), 179.1 (58), 191.1 (16), 135.1 (8) | 353.1→93.2, 127.2 (8), 111.1 (75) |
| 5-*O*-caffeoylquinic acid | 353.2 | 191.0 (100), 179.1 (5), 135.1 (1) | 353.2→85.2, 171.1 (45), 93.1 (42), 109.1 (40) |
| 1,3-di-*O*-caffeoylquinic acid | 515.0 | 353.0 (100), 335.2 (20) | 515.0→191.1, 179.0 (40), 135.1 (8) |
| 1,4-di-*O*-caffeoylquinic acid | 515.0 | 353.0 (100), 203.1 (10), 299.1 (10), 255.1 (4) | 515.0→173.0, 179.0 (60), 191.1 (4), 135.2 (15) |
| 1,5-di-*O*-caffeoylquinic acid | 515.2 | 353.0 (100), 335.1 (6), 191.0 (4) | 515.2→191.1, 179.0 (5) |
| 3,4-di-*O*-caffeoylquinic acid | 515.1 | 353.1 (100), 203.1 (10), 299.1 (10), 255.1 (4) | 515.1→173.1, 179.0 (69), 191.1 (49), 135.1 (10) |
| 3,5-di-*O*-caffeoylquinic acid | 515.0 | 353.1 (100) | 515.0→191.1, 179.0 (5), 135.1 (10) |
| 4,5-di-*O*-caffeoylquinic acid | 515.0 | 353.1 (100), 203.11 (12), 299.1 (10), 255.1 (5) | 515.0→173.1, 179.0 (60), 191.1 (40), 135.1 (6) |

[a] Identification according to Clifford et al., (2005)

The cinnamoyl esterase specificity of the three Lactobacillus strains is shown in Table 5.4 as percentage of monochlorogenic and dichlorogenic acids hydrolyzed and caffeic acid formed after 72 h of fermentation. The decrease in CQA is defined as the part of chlorogenic acid gone after 72 h of fermentation. The increase in CA was calculated based on the amount of product found after 72 h of fermentation. Due to the high reactivity of both compounds, the amounts of hydrolyzed CQA and formed CA were not completely consistent. The observed differences might be attributed to other reactions possibly taking place during fermentation, e.g., oxidation or decarboxylation. All three strains were able to hydrolyze all CQA isomers in a range of 0.11 mM – 1 mM (Figure 5.10). Formation of CA was not observed in any of the controls. A significantly higher hydrolysis rate by all Lactobacillus strains was observed for 5-CQA compared to the other monochlorogenic acids and dichlorogenic acids. After 72 h of fermentation, *L. helveticus* and *L. reuteri* completely hydrolyzed 5-CQA. The substrate decrease was accompanied by a corresponding amount of formed caffeic acid. *L. fermentum* displayed the lowest substrate consumption (60.4%). The present results are generally in agreement with results from literature (Sánchez-Maldonado et al., 2011). Guglielmetti et al. (2008) reported that some strains of *L. helveticus* were able to hydrolyze chlorogenic acid, but only one of two tested strains of *L. fermentum* was able to hydrolyze chlorogenic acid (41.6%). These authors also reported that the strains were able to hydrolyze ethyl ferulate and chlorogenic acid under the conditions chosen (1.4 mM for 72 h) in culture medium. Filannino et al. (2015) studied the changes in the phenolic acid profile during fermentation of broccoli and cherry juice by several lactobacilli species. The samples had a pH-value of 6.5 (broccoli juice) and 4.0 (cherry juice) and incubation was carried out at 37 °C for 24 h. The authors reported that only *L. reuteri* was able to hydrolyze chlorogenic acids under the conditions applied.

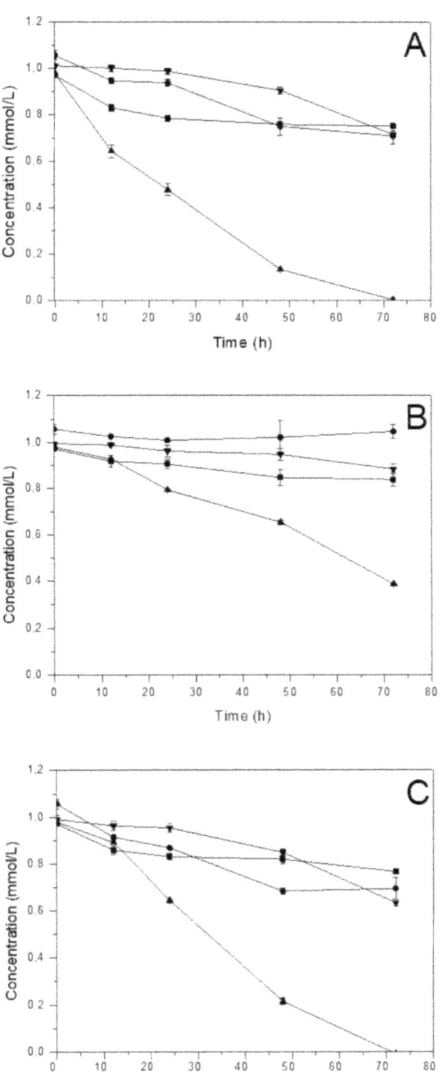


**Figure 5.10** Hydrolysis of 3-CQA (-■-), 4-CQA (-●-), 5-CQA (-▲-) and dichlorogenic acids (-▼-) by *L. reuteri* (A), *L. fermentum* (B) and *L. helveticus* (C) during 72 h fermentation. n=3

**Table 5.4.** Decrease of monochlorogenic acids and dichlorogenic acids and increase of caffeic acid during lactic acid bacteria fermentation in MRS medium after 72 h of incubation. [a, b, c]

| | *L. reuteri* | | *L. helveticus* | | *L. fermentum* | |
|---|---|---|---|---|---|---|
| **CQA** | **Decrease CQA %** | **Increase CA %** | **Decrease CQA %** | **Increase CA %** | **Decrease CQA %** | **Increase CA %** |
| **3-CQA** | 22.58 ± 1.09 [a,3] | 14.28 ± 3.33 [a,3] | 21.10 ± 0.79 [a,3] | 14.40 ± 1.42 [a,3] | 13.74 ± 3.05 [b,2] | 6.58 ± 1.55 [b,2] |
| **4-CQA** | 32.82 ± 4.45 [a,2] | 24.65 ± 1.01 [a,2] | 34.30 ± 4.24 [a,2] | 26.52 ± 3.71 [a,2] | n.d.* | n.d.* |
| **5-CQA** | 99.67 ± 0.86 [a,1] | 94.47 ± 3.95 [a,1] | 100 ± 0.45 [a,1] | 93.38 ± 6.48 [a,1] | 60.44 ± 0.24 [b,1] | 60 ± 5.12 [b,1] |
| **di-CQAs** | 27.70 ± 2.70 [b,2,3] | 20.14 ± 2.71 [b,2,3] | 36.03 ± 2.70 [a,2] | 30.89 ± 1.99 [a,2] | 10.77 ± 2.07 [c,2] | 4.41 ± 4.25 [c,2] |

[a] Values are mean ± standard deviation (n = 3); [b] Means in the same line with a different letter are significantly different ($p < 0.05$); [c] Means in the same column with a different number are significantly different ($p < 0.05$); *n.d., not detected

In contrast to Bel-Rhlid et al. (2013), who assessed the transformation of the sum of caffeoyl quinic acids from green coffee extracts by *L. johnsonii*, the present results deal with the hydrolysis of distinct monochlorogenic and dichlorogenic acid isomers and the corresponding formation of CA from the individual isomers. The results showed a strain variability in the cinnamoyl activity by these LAB. In the present study, 3-CQA decrease and hydrolysis rates of Lactobacillus strains were lowest in comparison with the other substrates. Highest cleavage rates of 3-CQA were observed for *L. reuteri* (22.6%), followed by *L. helveticus* (21.1%) and *L. fermentum* (13.7%). A maximum formation of caffeic acid from 3-CQA was found for *L. helveticus* (14.4%), followed by *L. reuteri* (14.3 %) and *L. fermentum* (6.6 %). The differences between the results for *L. helveticus* and *L. reuteri* were not significant.

Contrary to the reported mechanism of the hydrolysis of chlorogenic acids by LAB (Filannino et al., 2018), cleavage of 4-CQA followed a different pathway. While L. fermentum was not able to hydrolyze 4-CQA, incubation with *L. helveticus* and *L. reuteri* resulted in a decrease in 4-CQA and an increase in CA. But also 3-CQA and 5-CQA isomers were formed in considerable amounts in these fermentations. The isomers were not detected in the corresponding control samples. A second fermentation with 4-CQA was conducted to verify whether CA was a direct product of cleavage of 4-CQA or derived from other isomers formed initially. Formation kinetics of 3-CQA and 5-CQA as well as CA formation were studied during 6 hours; samples being taken at 0, 0.5, 1, 3, and 6 h. The results are shown in Figure 511. A decrease in 4-CQA was observed already after 30 minutes. Isomerization to 3-CQA and 5-CQA started within the first 30 minutes. Caffeic acid was also detected after 30 min, yet in lower concentrations. The sum of 3-CQA and 5-CQA isomers after 30 minutes corresponded well to the amount by which 4-CQA decreased. After 1 hour, concentration of all compounds remained relatively constant and only a slight decrease in 4-CQA and 5-CQA was observed, while CA was found to increase slowly after 30 minutes. Based on the aforementioned high hydrolysis rates by *L. reuteri* and *L. helveticus* observed after 72 h in the presence of 5-CQA, it can be suggested that CA formation from 4-CQA is mainly a result of hydrolysis of the initially formed 5-CQA. Accordingly, the higher concentration of CA after 6 h can be attributed to the respective significant decrease in 5-CQA rather than the hydrolysis of 3-CQA. The present findings suggest a possible isomerase activity exhibited by *L. reuteri* and *L. helveticus* prior to hydrolysis of the CQA into CA.

The cinnamoyl esterase from some lactobacilli and bifidobacteria was already identified, cloned, and characterized (Fritsch et al., 2017). These enzymes showed different responses regarding substrate affinity and reaction rate, depending on pH and temperature. However, among the broad group of microorganisms with the ability to hydrolyze chlorogenic acids, an isomerase activity has never been considered in the hydrolysis pathway. Lai et al. (2009) identified two esterase cinnamoyl enzymes (LJ0536 and LJ1228) in a probiotic *L. johnsonii*, showing the characteristic glycine-X-serine-X-glycine nucleophilic motif described previously for some cinnamoyl esterase which are able hydrolyze phenolic acids. Hydrolysis by this kind of enzymes is carried out by specific phenolic ester binding. Hydrolysis of substrates is possible by a sequence of two steps of acylation and deacylation. These two steps have been extensively explained by Lai et al. (2012). The active site of these esterase cinnamoyl enzymes is characterized by the triad of amino acids ser, his and asp, suggesting that these amino acids residues play an important role in the specificity of these enzymes action (Williamson et al., 1998; Lai et al., 2012). Deshpande et al. (2014) reported the hydrolysis and isomerization of CQA under acidic and basic conditions (pH 5 and 12). The authors reported a high pH dependency of the hydrolysis and the isomerization and concluded that the basic conditions are more favorable. Likewise, they suggested a mechanism for the isomerization between 3-CQA, 4-CQA and 5-CQA under basic conditions (Figure 5.12). The structures of the two different intermediates leading to either 5-CQA ore 3-CQA apparently differ in stability.

The equatorial orientation of the diester of the 5-CQA intermediate is favored over the axial orientation of the 3-CQA intermediate and, thus, 5-CQA is formed preferably. This is supported by the higher concentrations of 5-CQA at the beginning of the fermentation (figure 5.11), whereas the rather small differences may be explained by the faster hydrolysis of 5-CQA occurring concomitantly to the isomerization.

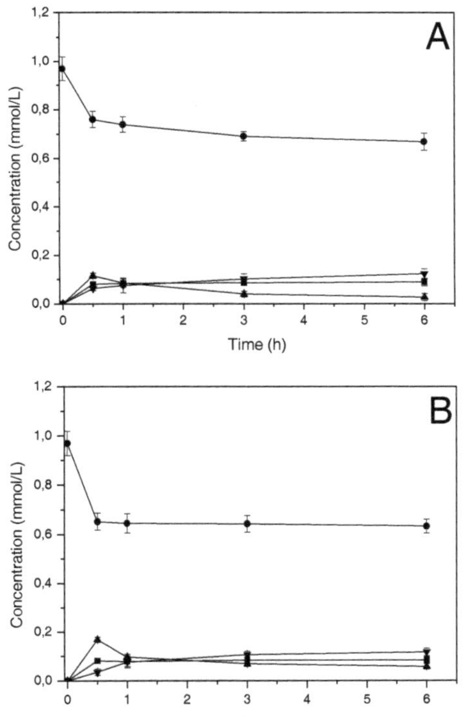


**Figure 5.11.** Formation of 3-CQA (-■-) and 5-CQA (-▲-) isomers and caffeic acid (-▼-) from 4-CQA (-●-) by *L. helveticus* (A) and *L. reuteri* (B) during 6 h of incubation. n=3

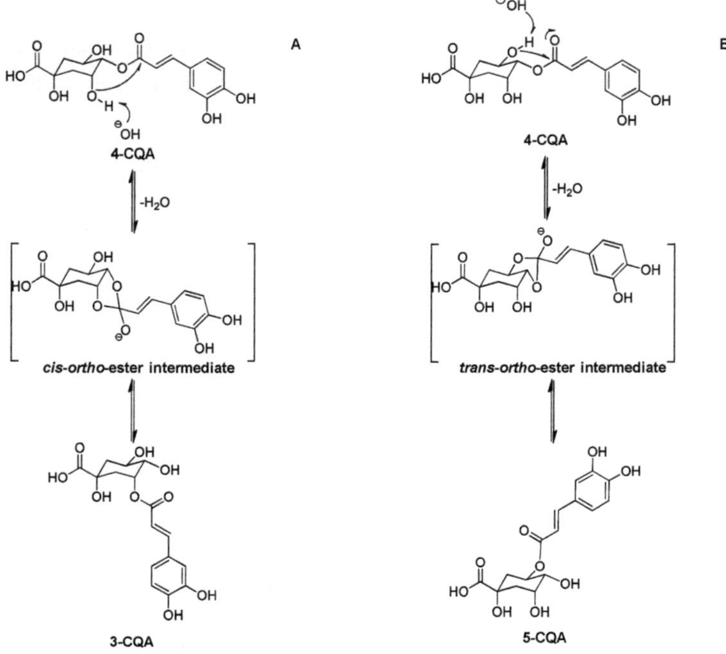


**Figure 5.12.** Mechanism of isomerization of 4-CQA according to Deshpande et al. (2014). Attack of C-3 to form 3-CQA (A) and attack of C-5 to form 5-CQA (B)

In the present study, a chemical hydrolysis can be excluded due to the absence of caffeic acid in the controls run for each fermentation. Regarding the results found for the fermentation of 4-CQA, it can be assumed that the cleavage of 4-CQA involves a sequence of two microbial activities. Initially, 4-CQA is isomerized into 3- and 5-CQA by a presumed isomerase activity and in the second reaction the already known hydrolase cleaves the formed 3-CQA and 5-CQA into CA and quinic acid. The mechanism of isomerization of 4-CQA by LAB prior to the hydrolysis can be explained according to the mechanism showed by Deshpande et al. (2014). This is the first study suggesting an isomerase and cinnamoyl esterase activities exhibited by Lactobacillus species in the metabolism of 4-CQA.

The ability of *L. fermentum*, *L. helveticus*, and *L. reuteri* to hydrolyze di-CQAs was also assessed and was found to be lower than the hydrolysis rate of 5-CQA (table 5.4). Microbial hydrolysis of CA resulting from the hydrolysis of dichlorogenic acids by LAB was confirmed by the absence of CA in all controls, where no traces were detected. *L. helveticus* reached the highest hydrolysis rate (36.1%) followed by *L. reuteri* (27.7%) and *L. fermentum* (10.8%). Contents of chlorogenic acids in Jerusalem artichoke have been determined by Kapusta et al. (2013), and 3,5- di-O-caffeoylquinic acid was found to be the major component. Other isomers identified in the present work are 1,4-di-O-caffeoylquinic acid, 3,4-di-O-caffeoylquinic acid, 4,5-di-O-caffeoylquinic acid, and cis-4,5-di-O-caffeoylquinic acid. *L. fermentum*, which did not show any activity to hydrolyze 4-CQA, exhibited the lowest degradation and hydrolysis rates for di-CQAs. These low rates can be attributed to the quantities of dichlorogenic acids with a caffeic acid bound at position 4. CA formation may be a product of hydrolysis of other dichlorogenic acids present. As a consequence, 4-CQA was still present after 72 h of fermentation. *L. helveticus* and *L. reuteri* led to higher hydrolysis rates and formation of CA than *L. fermentum*. These results are consistent with the higher affinity of the strains to 5-CQA and 3-CQA as well the new and suggested pathway for the degradation of 4-CQA.

Factors influencing the induction of the production of these microbial enzymes involved in the hydrolysis of chlorogenic acids are not well established. Further investigation of these factors is necessary since they affect enzyme properties such as substrate specificity, reaction velocity, and pH- and temperature dependency. More information is needed on the mechanism of induction of the presumed isomerase enzymes involved in the cleavage of 4-CQA and their isolation and full characterization.

## 5.4. Synthesis of pyranoanthocyanins

Since their identification in wine, pyranoanthocyanins are of interest due to their higher stability in comparison with anthocyanins. Despite the great advantage presented by their higher stability, the difficulties caused by the low concentrations in which they are formed, their characterization and isolation is challenging. In this study, LAB possessing CD activity and CE activity were used in co-fermentations in the production of a food coloring source rich in pyranoanthocyanins applicable in the food industry.

### 5.4.1. Formation of pyranoanthocyanins in MRS media

MRS media and buffer solutions containing 4-VP, 4-VG and 4-VC products of *L. sakei* CD activity (see section 4.3.1) were supplemented with purified blackberry anthocyanins and used for synthesis anthocyanins-adducts.

For this purpose, blackberry anthocyanins were isolated, and their identification was based on literature (Dugo et al., 2001; Fan-Chiang & Wrolstad, 2005; Zhang et al., 2012). Figure 5.13 shows the four anthocyanins identified in blackberries extract, all of them cyanidin derivatives and Cy-3-glu found in major concentration.

After 5 days of reaction in MRS media and buffer solution two anthocyanins-adducts were synthetized. In the control samples, no anthocyanin-adducts were formed, however, a decrease in anthocyanins concentration was observed resulting from thermal degradation. Figure 5.14 shows Cy-3-glu and the two hydroxyphenyl-pyranoanthocyanins formed during reaction, i.e., Pyr1 and Pyr2. Pyr1 resulted in being an adduct of Cy-3-glu while Pyr2 resulted from another anthocyanin present in a minor concentration, i.e., cyanidin-3-(6″-malonyl) glucoside or cyanidin-3-dioxaloylglucoside and the vinyl derivatives.

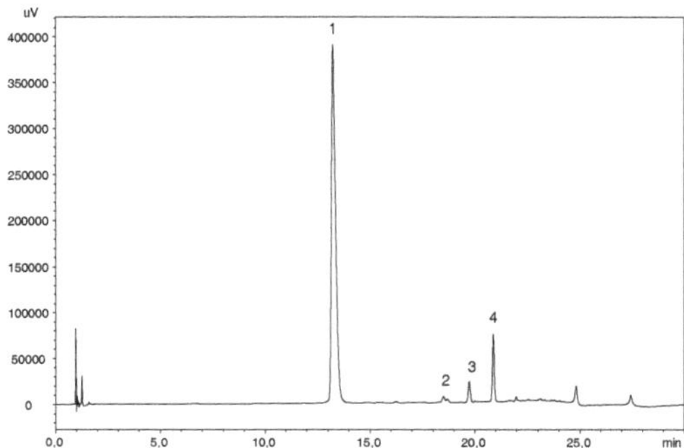


**Figure 5.13.** UHPLC separation of blackberry anthocyanins cyanidin-3-glucoside (1), cyanidin-3-xyloside (2), cyanidin-3-(6″-malonyl) glucoside (3), and cyanidin-3-dioxaloylglucoside (4). Chromatogram recorded at 520 nm

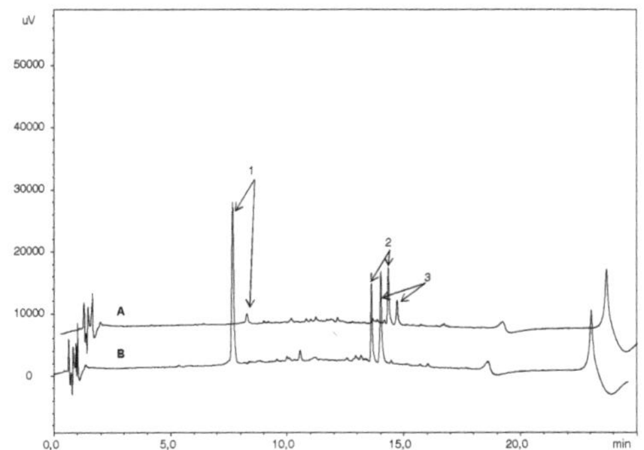


**Figure 5.14.** Chromatogram of the synthesis of pyranoanthocyanins in MRS medium (A) and buffer solution (B) by the reaction of CA at pH 4.5. Cy-3-glu (1) and pyranoanthocyanins fractions (2 and 3) recorded at 520 nm

From a previous section (see section 5.1.2) results obtained after CD activity from HCA by *L. sakei* were extracted and presented in table 5.5 since they were needed for the understanding of this study.

Results are at two pH-value and expressed in mM $L^{-1}$ which represent the concentration of the vinyl derivatives at the initial reaction time for pyranoanthocyanins synthesis. Analysis of samples after 24 h of reaction revealed the formation of Pyr1. Concentrations of all vinyl derivatives were higher in media than in buffer solution, however, synthesis of Pyr1 was favored in buffer solution. This might be due to all compounds conforming to the MRS media affecting anthocyanins and pyranoanthocyanins stability. This behavior was also previously seen during CD activity (see section 5.1.2).

**Table 5.5.** Concentration of vinyl derivatives at initial reaction time and pyranoanthocyanin formed after 24 h of reaction [a]

| | pH 4.6 | | pH 5.6 | |
|---|---|---|---|---|
| **Media** | **4-VP** | **Pyr1** | **4-VP** | **Pyr1** |
| **MRS** | 4.70±1.55 | 0.03±0.00 | 4.49±0.32 | 0.03±0.00 |
| **Buffer** | 0.10±2.52 | 0.07±0.01 | 0.60±4.74 | 0.04±0.01 |
| **Media** | **4-VG** | **Pyr 1** | **4-VG** | **Pyr 1** |
| **Medium** | 0.19±4.23 | 0.04±0.01 | 0.16±0.98 | 0.04±0.00 |
| **Buffer** | 0.19±0.72 | 0.02±0.00 | 0.17±4.01 | 0.02±0.00 |
| **Media** | **4-VC** | **Pyr 1** | **4-VC** | **Pyr 1** |
| **Medium** | 2.58±4.58 | 0.03±0.00 | 2.52±0.02 | 0.03±0.00 |
| **Buffer** | 0.46±3.88 | 0.04±0.00 | 1.60±7.59 | 0.05±0.01 |

[a] Results are expressed as mM $L^{-1}$; Pyr 1 is Cy-3-glu-4-VP for 4-VP, is Cy-3-glu-4-VG for 4-VG, and is Cy-3-glu-4-VC for 4-VC

Conversion rate of anthocyanins into hydroxyphenyl-pyranoanthocyanins are shown in table 5.6. As expected, higher degradation of anthocyanins in buffer and media solution resulted in samples at the highest pH-value assessed (5.6). The well-documented pH-dependent changes in anthocyanins structure, may explain the lower pyranoanthocyanin formation rate at the highest pH assessed (5.6) in comparison to the lowest pH-value tested (4.6). An exception was found with samples containing 4-VC, where the highest formation of pyranoanthocyanins was at pH 5.6. Moreover, at both pH-values among all three vinyl derivatives, results showed that pyranoanthocyanin formation was favored with 4-VP toward with 4-VC or 4-VG. This can be demonstrated with the formation rate of Pyr1 at pH 4.6 in buffer solution (table 5.6). Pyr1 with 4-VP resulted 5.05% higher than with 4-VG and 7.04% higher than 4-VC.

Hydroxyphenyl-pyranoanthocyanins are one type of pyranoanthocyanins found in aged wines (Schwarz et al., 2003c). In wine, these pigments are the result of the reaction of anthocyanins with HCA or their vinyl derivatives, the latter resulting from enzymatic microbial activities from HCA. Reports suggest that copigmentation reactions are more efficient at a high copigment ratio (Malien-Aubert et al., 2001). This phenomenon has been suggested to be the first step in the formation of pyranoanthocyanins (Maccarone et al., 1985; Lee & Nagy, 1990). It has been reported that the new pyran ring provides higher color intensity and more stability than the natives anthocyanins. Therefore, these compounds may be good candidates to be used as natural colorants. However, formation of this compounds occurs slowly (Rentzsch et al., 2007). Schwarz et al. (2003) reported a one-step reaction for the synthesis of hydroxyphenyl-pyranoanthocyanins. These authors applied a model solution for the synthesis of pyranoanthocyanins using malvidin-3-glucoside and 4-VC and suggesting that high quantities of free HCA could facilitate the formation of pyranoanthocyanins.

**Table 5.6.** Anthocyanin conversion rate (%) into pyranoanthocyanins in MRS media and buffer solution after 5 days of reaction

| | | pH 4.6 | | | pH 5.6 | | |
|---|---|---|---|---|---|---|---|
| | **Media** | **Pyr 1** | **Pyr 2** | **Cy-3-glu** | **Pyr 1** | **Pyr 2** | **Cy-3-glu** |
| **4-VP** | **MRS** | 5.40±0.30 | 3.03±0.02 | 2.76±0.52 | 5.14±0.52 | 3.47±-0.97 | 3.24±0.21 |
| | **Buffer** | 13.67±2.63 | 3.99±0.30 | 46.05±1.97 | 8.49±1.97 | 4.34±0.11 | 5.51±0.15 |
| **4-VG** | **MRS** | 8.44±1.28 | 5.01±0.64 | 12.02±3.72 | 7.84±0.28 | 5.02±0.36 | 9.06±0.74 |
| | **Buffer** | 8.62±0.63 | 2.64±0.07 | 73.84±3.48 | 3.02±0.61 | 1.73±0.13 | 21.30±0.26 |
| **4-VC** | **MRS** | 4.96±0.00 | 3.15±0.10 | 1.96±0.14 | 5.35±0.05 | 3.42±0.05 | 2.30±0.17 |
| | **Buffer** | 6.63±0.41 | 8.18±0.38 | 9.83±1.08 | 9.85±2.58 | 4.38±0.20 | 3.14±0.02 |

In agreement with these authors, this work reports the formation of pyranoanthocyanins by one-step reaction with 4-VP, 4-VC, or 4-VG. After 5 days of reaction, a maximum formation rate of 13.67% from the reaction with 4-VP was reached. Some authors reported the first detection of pyranoanthocyanins after 5 days of reaction (Schwarz et al., 2003c; He et al., 2010a). Additionally, pyranoanthocyanin formation has been reported as a slow reaction (Schwarz & Winterhalter., 2003; He et al., 2010a; de Freitas & Mateus, 2011). Since anthocyanins are more stable at acidic conditions, the higher conversion rate obtained at pH 4.6 was expected since degradation rate at this pH was slower compared to pH 5.6. However, more reaction time was probably needed to complete the anthocyanins conversion in this study. Moreover, previous works have suggested that the amount of the precursor reacting with anthocyanins highly influenced the formation of pyranoanthocyanins (Romero et al., 1999; Romero & Bakker, 2000; Schwarz et al., 2003; Morata et al., 2003; Garcia-Alonso et al., 2005). Therefore, vinyl derivatives might not have been present in enough quantity to compete against anthocyanin degradation through the formation of pigment/copigment complexes. Degradation of anthocyanins mainly occurs due to oxidation and cleavages of covalent bonds which is accelerated by high temperatures (Zhang et al., 2012). This study carried out at 37 C showed anthocyanins' degradation which was corroborated by the brownish color exhibited by the control samples at the end of the reaction time. In samples, even at the low concentration in which hydroxyphenyl-pyranoanthocyanins were synthetized, the impact on the color was visible.

Figure 5.9 shows the color presented by the media at the initial reaction time (A) and the color presented by the media after 5 days of reaction (B) which is the typical color presented by hydroxyphenyl-pyranoanthocyanins. These pigments have a maximum absorption wavelength between 495 and 520 nm with a characteristic absorption peak in the 420 nm region (Fulcrand et al., 1998; Morata et al., 2006; He et al., 2010c; Marquez et al., 2013).

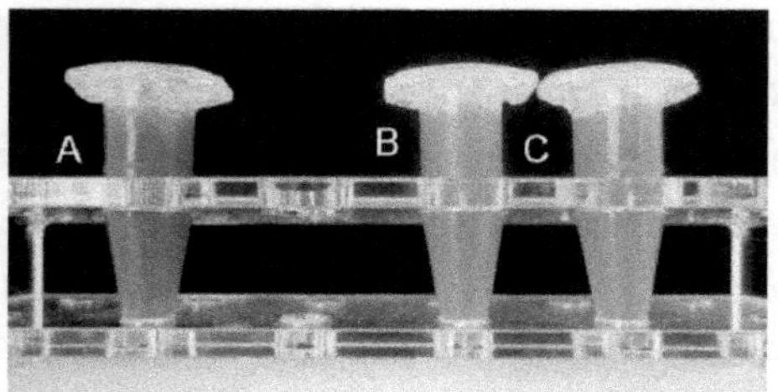


**Figure 5.9.** MRS media supplemented with Cy-3-glu (A) at initial reaction time and MRS media after 5 days of reaction with 4-VP (B), and 4-VG (C)

### 5.4.2. Effect of the media on the stability of pyranoanthocyanins

Cy-3-glu-4-VP and Cy-3-glu-4-VG, previously synthetized (see section 5.4.1), and their precursor Cy-3-glu were analyzed regarding their stability at fermentation conditions.

In figure 5.10, the kinetics of Cy-3-glu degradation presented in media and in buffer solutions at two different pH-values are shown. Cy-3-glu showed greater stability at pH-value of 4.6. At this pH-value Cy-3-glu degradation was found to be 20% greater in MRS media than in buffer. At the end of the trial, quantification showed 33.35% and 52.44% of Cy-3-glu remaining in MRS media and buffer solution, respectively. In contrast, at pH-value of 6.6, around 60% of Cy-3-glu was degraded within the first 24 h and disappeared almost completely during the second day of study.

It is well-known that anthocyanins degradation follows a first order reaction kinetic (Adams, 1973; Romero & Bakker, 2000), this could be demonstrated by trials at pH 4.6. Nevertheless, at pH 6.6 the faster degradation did not allow their first order kinetic modelling.

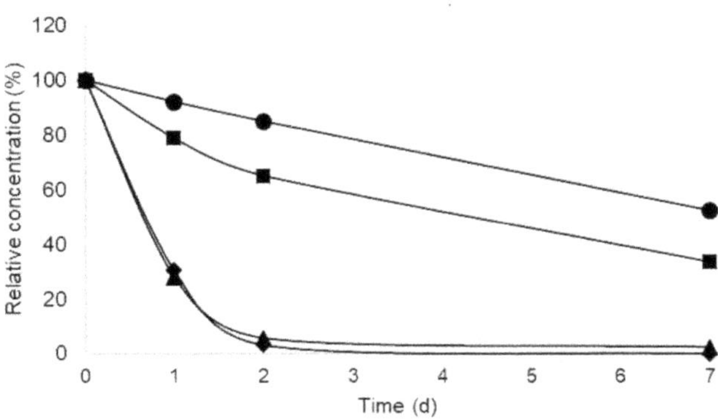


**Figure 5.10.** Degradation kinetic of Cy-3-glu in buffer solution (●) and MRS media (■) at pH 4.6, and in buffer solution (♦) and MRS medium (▲) at pH 6.6

The first order reaction rate constant (*k*) and the half-life (t ½) were calculated for trials conducted at pH 4.6 by the following 5.1 and 5.2 equations, respectively.

$$\ln (C/Co) = -kt \quad (5.1)$$

$$t_{1/2} = ln\, 2/\, k \quad (5.2)$$

where co is the initial Cy-3-glu concentration, C corresponding to anthocyanins content after storage time (t). The total anthocyanins degradation applying this first order reaction is shown in figure 5.11. The half-life values calculated were 4.6 days and 7.4 days in MRS media and buffer solution, respectively.

Figure 5.12 shows the kinetics of Cy-3-glu-4-VP degradation. Samples at pH 4.6 in buffer solution showed no losses in Cy-3-glu-4-VP concentration, whereas 38.4% losses in MRS media were presented. However, at pH-value of 6.6 losses of 55.0% and 33.4% in MRS media and buffer solution were presented, respectively.

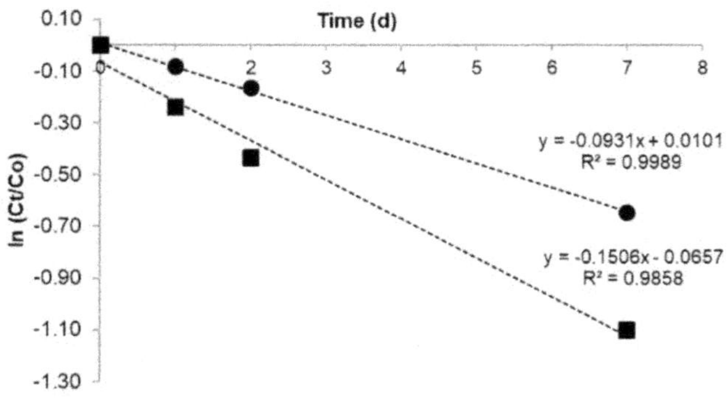


**Figure 5.11.** First order kinetic for retention of Cy-3-glu in buffer (•) and MRS medium (▪) at pH 4.6

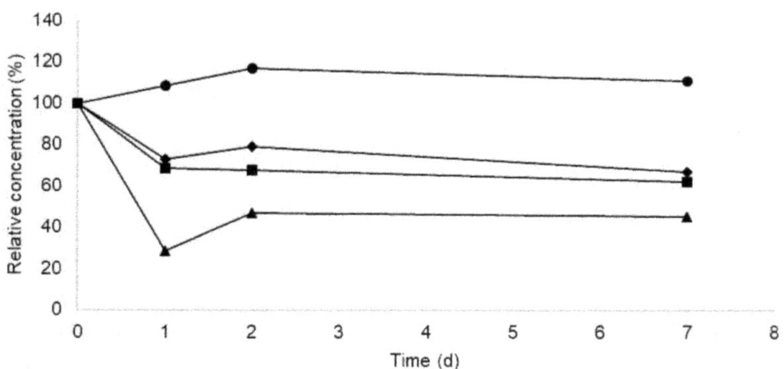


**Figure 5.12.** Degradation of 10-hydroxyphenyl-pyranocy-3-glu in buffer solution (•) and MRS medium (▪) at pH 4.6 and in buffer solution (♦) and MRS media (▲) at pH 6.6

The degradation rate followed by Cy-3-glu-4-VG at pH-value of 4.6 resulted in 47.6% and 54.10% in buffer and MRS media, respectively. However, samples at pH 6.6 remained without losses throughout the period of study (figure 5.13).

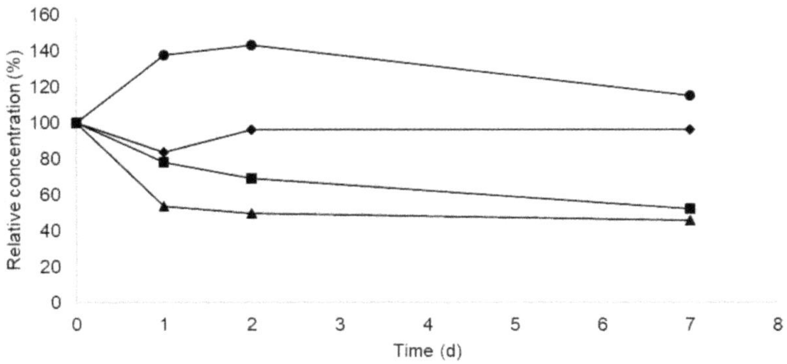


**Figure 5.13.** Degradation of 10-guacyl-pyranocy-3-glu in buffer solution (●) and MRS media (■) at pH 4.6 and in buffer solution (♦) and MRS media (▲) at pH 6.6

In general, the stability test showed that, the main advantage presented by the two hydroxyphenyl-pyranoanthocyanins, Cy-3-glu-4-VP and Cy-3-glu-4-VG with their Cy-3-glu precursor is their greater stability in both media and pH-values assessed. It is well-known that the color exhibited by anthocyanins is strongly affected by the pH-value of the solution. Anthocyanins are highly stable at acidic conditions, whereas an increase in the pH-value leads to the formation of colorless carbinol bases compounds. Among pyranoanthocyanins, most of the research has been done in wine. These works reported pyranoanthocyanins adducts to be more stable than the genuine anthocyanin (Romero & Bakker, 2000; Asentorfer et al., 2001; Schwarz et al., 2003; Schwarz & Winterhalter, 2003).

The flavylium cation of the anthocyanins is prone to nucleophilic attack due to electron deficiency, whereas hydroxyphenyl-pyranoanthocyanins formation involves the nucleophilic addition of the phenol group between C-4 and C-5 followed by oxidation and dehydration reactions.

This new pyran ring shields the molecule against hydration causing an increase in the stability (Malien-Aubert et al., 2001).

In this study, all three pigments were very susceptible to degradation in MRS media. The lower stability obtained in MRS media may be attributed to the media composition that might contribute to their faster degradation. Stability analysis of Cy-3-glu can explain previous results (see section 5.4). It was possible to show that Cy-3-glu degradation occurs at a faster rate at pH 6.6 independently of the media. It might be that degradation of Cy-3-glu occurs faster than the time needed for pyranoanthocyanins formation which explains the higher amount of pyranoanthocyanins at pH 4.4 in buffer solution.

Romero and Bakker (2000) evaluated malvidin-3-glucoside degradation and visitin A formation with pyruvic acid at 10 °C, 15, °C 20 °C and 32 °C. These authors reported that, as expected, anthocyanins decrease faster at higher temperatures following the first-order kinetics, however, samples containing pyruvic acid, showed that, at low temperatures, the degradation rate of malvidin-3-glucoside may be accelerated, while, at higher temperatures, the opposite effect was observed.

### 5.4.3. Effect of the pH value in the decarboxylation activity by LAB in black carrot juice

Black carrot juice is a source rich in phenolic acids, including *p*-CA, FA, and CA, however, 5-CQA has been reported as the predominant HCA (Kammerer et al., 2004a; Schwarz et al., 2004). Phenolic acids have been reported toxic for some LAB, therefore, decarboxylase activity of *L. plantarum*, *L. mali* and *L. sakei* was assessed. After the supplementation of CA in equal quantity in which 5-CQA concentration is found in juice, CD activity of LAB was assessed at two different pH values (5.6 and 4.6). In this study, the formation of pyranoanthocyanins was directly related to decarboxylation activity by LAB. 5-CQA it is reported to be predominant in black carrot, for this reason, 5-CQA was quantified.

Additionally, supplementation of CA in BCJ with 5-CQA was done because the generation of hydroxyphenyl-pyranoanthocyanins is carried out in a ratio 1:1 between anthocyanins and vinyl derivatives.

Black carrot anthocyanins could be identified by UHPLC-MS analysis and comparison with literature. Compounds identified matched with the fragmentation behavior of those reported in the literature (Kammerer et al., 2004a; 2004b). Table 5.7 shows ten anthocyanins (1-10) and traces of three pyranoanthocyanins (12, 14, and 16) that could be identified in BCJ.

Black carrot anthocyanins were acylated anthocyanins with HCA or hydroxybenzoic acids. Anthocyanins based on a cyanidin-aglycon contributed to the major fraction in BCJ. Regarding all anthocyanins, cyanidin 3-xylosyl (feruloylglucosyl) galactoside (6) characteristic of black carrots was found in large quantities followed by cyanidin 3-xylosyl(glucosyl)galactoside (1), cyanidin 3-xylosylgalactoside (2), and cyanidin 3-xylosyl (sinapoylglucosyl) galactoside (4).

After 48 h, no CA could be detected in fermented samples at the two pH-values assessed, suggesting a decarboxylation rate of 100%. Additionally, a decrease in 5-CQA content was observed in fermented, and control samples, but no 4-VC was detected. A previous characterization of *L. plantarum*, *L. mali*, and *L. sakei* used in this work, showed that these strains were not able to hydrolyze CQA. Therefore, this decrease in 5-CQA concentration could be attributable to the thermal degradation (37 °C).

In non-fermented juice used as a control, the concentration of pyranoanthocyanins was found to be below the limit of quantification. In fermented-juice, decarboxylation of CA by LAB strains led to the formation of 4-VC and 4-VG substituted pyranoanthocyanins.

**Table 5.7.** Anthocyanins and pyranoanthocyanins identified in BCJ before and after fermentation [a]

| | | *m/z* | | |
|---|---|---|---|---|
| Peak | R.T (min) | $[M]^+$ | $MS^2$ | Compound |
| 1 | 11.55 | 743 | 287.01 | cyanidin 3-xylosyl(glucosyl)galactoside |
| 2 | 12.59 | 581 | 287.01 | cyanidin 3-xylosylgalactoside |
| 3 | 13.22 | 863 | 287.04 | cyanidin 3-xylosyl(*p*-hydroxybenzoylglucosyl)galactoside |
| 4 | 13.54 | 949 | 287.02 | cyanidin 3-xylosyl(sinapoylglucosyl)galactoside |
| 5 | 13.37 | 595 | 301.04 | peonidin 3-xylosylgalactoside |
| 6 | 13.81 | 919 | 287.02 | cyanidin 3-xylosyl(feruloylglucosyl)galactoside |
| 7 | 13.96 | 889 | 287.03 | cyanidin 3-xylosyl(coumaroylglucosyl)galactoside |
| 8 | 14.18 | 963 | 301.08 | peonidin 3-xylosyl(sinapoylglucosyl)galactoside |
| 9 | 14.40 | 903 | 271.03 | pelargonidin 3-xylosyl(feruloylglucosyl)galactoside |
| 10 | 14.48 | 933 | 301.02 | peonidin 3-xylosyl(feruloylglucosyl)galactoside |
| 11 | 16.42 | 875 | 419 | Vinylcatechol adduct of 1 |
| 12 | 17.29 | 713 | 419.08 | Vinylcatechol adduct of 2 |
| 13 | 17.90 | 697 | 403 | Vinylphenol adduct of 2 |
| 14 | 18.18 | 1051 | 419.05 | Vinylcatechol adduct of 6 |
| 15 | 18.10 | 1081 | 419 | Vinylcatechol adduct of 4 |
| 16 | 18.20 | 727 | 433.11 | Vinylguaiacol adduct of 2 |

[a] Identification according to Cuevas Montilla et al., (2011)

This could be confirmed by the generation of fragments with *m/z* 419 and *m/z* 433 showing that all pyranoanthocyanins were 4-VC and 4-VG adducts, respectively. HPLC-ESI-$MS^n$ analysis of fermented-juice revealed four pyranoanthocyanins with *m/z* 875, 713, 1051, and 727 corresponding to compounds 11,12,14 and 16 presented in table 5.7, respectively. Schwarz et al. (2004) identified these compounds in BCJ, however, due to the low concentration they were found could not be quantified.

As reported by Kammerer et al. (2004a), black carrots are a rich source of phenolic acids, i.e., 5-CQA, CA, *p*-CA, and FA. Black carrots have been reported to contain 5-fold more phenolic acids than yellow, orange, or white carrots (Alasalvar et al., 2001). In this analysis, 4-VC and 4-VG adducts detected in fermented-juice arose from CD activity displayed by the LAB strains, which has been previously discussed in section 5.1. Total decarboxylation of CA in this trial by LAB strains, agreed with the previous results obtained in section 5.1.1. Detection of 4-VC after fermentation was not possible, probably because this compound is highly reactive and could react with many other compounds in BCJ. Morata et al. (2007) reported that vinyl derivatives can easily react with anthocyanins to form more stable compounds. As reported in the literature, higher quantities of the compound reacting with anthocyanins could result in higher pyranoanthocyanins formation (Malien-Aubert et al., 2001). The metabolism of FA by these LAB strains was found to occurs at a lower rate compared to CA (Rodríguez et al., 2008; Buron et al., 2011). Thereby, the formation of adducts of 4-VC may largely be explained. From all pyranoanthocyanins identified, the one possessing a m/z 875 formed by the reaction of cyanidin 3-xylosyl (glucosyl) galactoside (1) with 4-VC was found as the major product after fermentation. Cyanidin 3-xylosyl(feruloylglucosyl) galactoside (6) was the main anthocyanin found in non-fermented juice and, it has been reported to be present in black carrot in concentrations ranging from 40 to 80%.

Despite the higher concentration of compound 6, adduct 11 was the predominant product which was originated from compound 1 with 4-VC. However, adduct 12 was the second majoritarian product which was originated from compound 6 after cleavage of ferulic acid (Figure 5.13) or from compound 2.

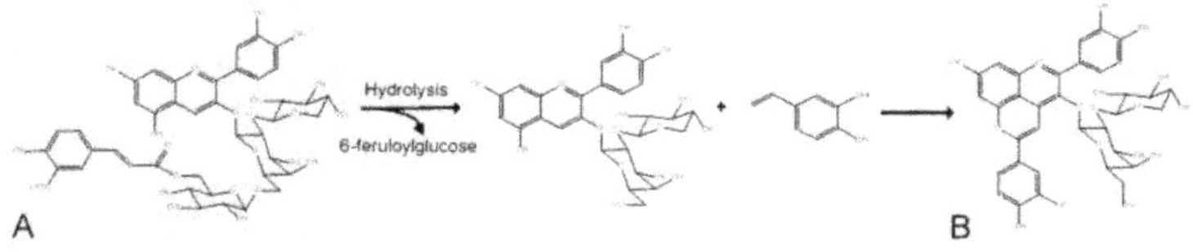

**Figure 5.13** Anthocyanin 6 (A) participating in the synthesis of pyranoanthocyanin 12 (B)

Results present in Figure 5.14 shows the anthocyanins conversion rate after 48 h at both pH-values assessed. The highest conversion rate of anthocyanins into pyranoanthocyanins was obtained with *L. sakei* (55%) at pH 4.6. Samples taken after 62 d showed a slight increase in pyranoanthocyanins concentration with the maximum conversion rate of 59% by *L. sakei* (pH 4.6).

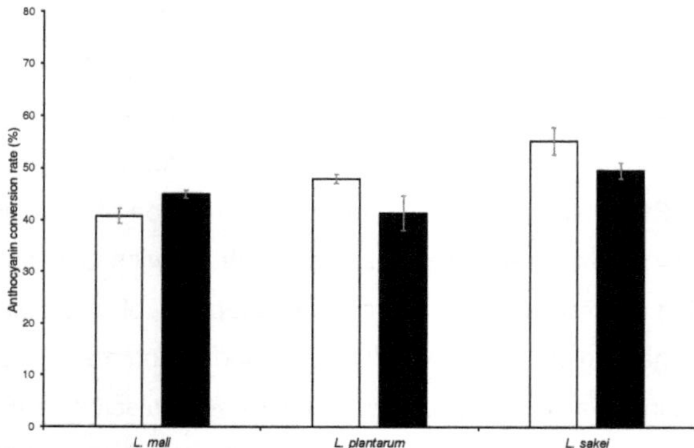


**Figure 5.14.** Anthocyanins conversion rates (%) of black carrot juice fermented by lactobacilli at pH 4.6 (white bars), and 5.6 (black bars)

It could be suggested that the anthocyanin-adducts are resulting from the reaction with HCA. As previous authors reported, pyranoanthocyanins formation through condensation with vinyl derivatives proceeds faster than the reaction with HCA and anthocyanins (Wang et al., 2003; Schwarz et al., 2003; de Freitas & Mateus, 2011).

As previously reported, black carrot juice anthocyanins have been reported highly stable at a wide range of pH and temperature (Türker & Erdoğdu, 2006; Sadilova et al., 2006; Kirca et al., 2007; Khandare et al., 20011; Türkylmaz & Özkan, 2012; Shukla & Vankar, 2013). However, higher rates of conversion were found at the lowest pH, as previously discussed (section 5.5.1), faster degradation occurred at this study conditions (37 C) and the higher pH-value (5.6). Figure 5.15 shows black carrot juice at the initial time and the orange tonalities typically presented by pyranoanthocyanins after fermentation.

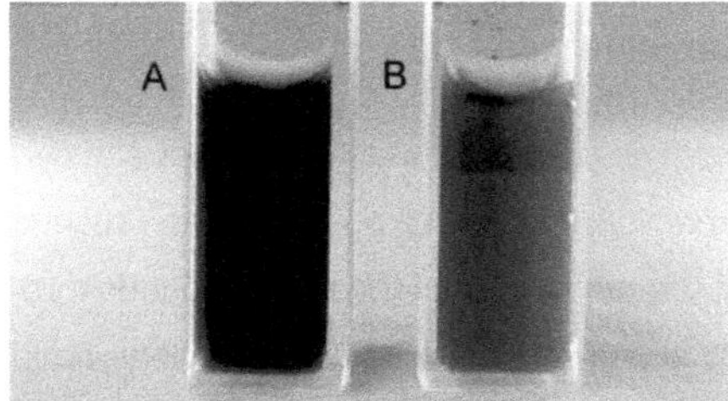


**Figure 5.14.** Samples of black carrot juice not-fermented (A) and fermented (B)

Concentrates of BCJ have been widely used as natural colorants (Yilmaz & Bilek, 2017). Kirca et al (2007) used black carrot concentrate to color some fruit juices and nectars and study black carrot anthocyanins. These authors reported anthocyanin degradation occurring fast at 37 °C. However, Kammerer et al. (2007), who used black carrots to impart color to canned strawberries, reported that the color of canned strawberries containing BCJ was highly improved while samples without added juice were reduced to 85%. Some authors attributed this result to the presence of acylated anthocyanins and their higher stability during processing and storage (Kammerer et al., 2007; Khandare et al., 2011). Acylated anthocyanins in black carrots have been reported to constitute up to 55 and 99% of the total content. Stability exhibited by acylated anthocyanins has been attributed to the intramolecular copigmentation reactions preventing nucleophilic attack of water leading to colorless compounds (Malien-Aubert et al., 2001). In addition to this higher stability, Schwarz et al. (2004), reported the increase of pyranoanthocyanins during BCJ storage formed through the reaction between anthocyanins and HCA. It has been reported that CQA is unable to react with anthocyanins (Schwarz et al., 2003c; Schwarz et al., 2004), additionally, reaction with HCA is reported to proceed very slowly. Therefore, and based on the results presented in this study, fermentation by LAB contribute to the formulation of a coloring foodstuffs rich in stable compounds as are pyranoanthocyanins.

### 5.4.4. Synthesis of pyranoanthocyanins through co-fermentation with LAB

Schwarz et al. (2003) showed the synthesis of pyranoanthocyanins by a one-step reaction in a model solution. The present study showed a new scheme for the generation of these compounds through fermentation by LAB from an accessible source, BCJ. Previous chapters showed the higher activity of *L. plantarum*, *L. mali* and *L. sakei* to CD activity and *L. reuteri*, *L. helveticus* and *L. fermentum* showing higher CE activity (Sections 5.1 and 5.3). Synthesis of pyranoanthocyanins in BCJ took place through three different reactions: first the hydrolysis of esters to release HCA, second their decarboxylation to produce their vinyl derivatives, and finally the reaction of anthocyanins with these vinylphenols.

During a first trial, juice was fermented by binary combinations comprising *L. reuteri* with *L. plantarum*, *L. mali*, or *L. sakei*. After fermentation, juice showed a decrease in anthocyanins concentration, nevertheless, no pyranoanthocyanins were found. Besides this result, CA was detected from CE activity displayed by *L. reuteri*, however, LAB strains possessing CD activity could not grow. Based on the previous high CE activity reported by *L. reuteri* (section 5,3), it was expected that a higher rate of hydrolysis could lead to high 4-VC generation, however, since *L. plantarum*, *L. mali* and *L. sakei* did not grow, no decarboxylation products were detected. As reported by Kammerer et al. (2004a), BCJ contain high amounts of phenolic acids, which could also influence the bacterial growth of *L. plantarum*, *L. mali* and *L. sakei*. Another important factor affecting the enzymatic activities, could have been the pH variability between the two groups of LAB assessed. As it was reported in a previous section (see section 5.3.1), CE activity by LAB showed higher yields at a pH of 6.6 which was the initial pH-value presented by BCJ. In contrast, CD activities were favored at pH-values below 6.6 (see section 5.1). Therefore, in this study, all conditions were appropriate for CE activity by *L. reuteri*, showing, in the end, the dominance over LAB with CD activity. Considering these results, a second assay was developed in order to avoid the competition between these two groups of LAB which is shown in the diagram found in section 4.3.14. In a first stage, *L. reuteri* was inoculated in BCJ (J1) for hydrolysis of 5-CQA yielding available CA which could be further used by *L. plantarum*, *L. mali* or *L.* sakei for the generation of vinylcatechol.

After fermentation, a high decrease in 5-CQA concentration was observed in non-fermented juice used as control. The most likely reason is thermal degradation. This can be suggested since control juice was exposed in its totality to the temperature used in this study (37 °C). For fermented juice, the new trial allowed to avoid the high 5-CQA hydrolysis and thermal degradation of anthocyanins by several steps that were carried out. This study showed that compared with naturally occurring concentrations found in conventionally processed fruit products, fermentation with LAB revealed high conversion rates. Figure 5.15 shows the conversion of anthocyanin into pyranoanthocyanins. As it is possible to observe, among all binary combinations, the higher anthocyanins conversion rate was found by *L. reuteri* and *L. sakei* showing a 37% and 47% after 96 h and 31 d, respectively.

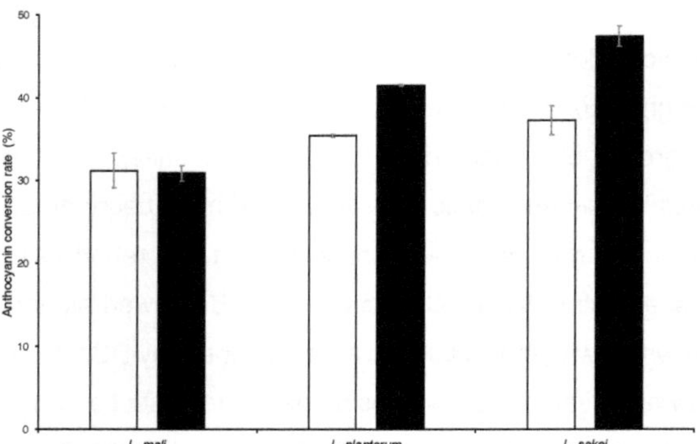


**Figure 5.15.** Anthocyanins conversion rates (%) of black carrot juice fermented by *L. reuteri* and *L. plantarum*, *L. mali*, or *L. sakei* samples taken after 96 h (white bars), and 31 d (black bar) of fermentation

Results showed that in a first stage (J1) *L. reuteri* showed a 5-CQA hydrolysis rate of 9%. Additionally, *L. reuteri* displayed the metabolism of CA by CD activity reaching a 61% conversion into 4-VC. After 24 h of fermentation, the supernatant was taken and supplemented with fresh juice to be posteriorly inoculated with *L. plantarum*, *L. mali*, or *L. sakei*, respectively. Results showed that from this supernatant after 24 h of fermentation, no decarboxylation products could be quantified. However, in the same period fresh juice was provided to *L. reuteri* (J2) to continue the hydrolysis into CA, however, once again this LAB strain was shown to favor the decarboxylation into 4-VC (71%) over the cleavage of 5-CQA. Finally, a third phase consisted in adding the supernatants resulting from previous steps J1 and J2 and add the rest of the fresh juice to complete the fermentation. At the end of the fermentation, contrary to the expected decrease of CA, results showed an increase in the concentration, with this increase attributable to thermal hydrolysis. This result could be confirmed by the decrease in 5-CQA concentration in non-fermented juice. Due to the lack of products from CD activity by *L. plantarum*, *L. mali*, or *L. sakei*, it can be suggested that 4-VC contributing in the generation of anthocyanins-adducts was most likely product of CD activity by *L. reuteri*. *L. reuteri* was chosen for this study due to the high yields while hydrolyzing 5-CQA (see section 5.3.1). Initially, BCJ showed a pH-value of 6.6 which, after 24 h, decreased to 3.7. The presence of lactic acid, as product of the pH reduction by *L reuteri* growth probably suppressed the growth of *L. plantarum*, *L. mali* and *L. sakei*. Additionally, a bacteriocin produced by *L. reuteri* could be responsible for *L. plantarum*, *L. mali* and *L. sakei* growth inhibition. This antimicrobial protein called reuterin has been previously reported by several authors (Toba et al., 1991; Kawai et al., 2001; Spinler et al., 2008). As described by Toba et al. (1991), the bacteriocin produced by *L. reuteri* has been shown to possess a broad inhibitory spectrum able to inhibit *Lactobacillus* strains, *Bacillus subtilis*, *Staphylococcus aureus*, among others (Gänzle et al., 2000). The formation of this bacteriocin could also explain the results found during the first trial. However, in previous sections, either CD activity or the production of a microbial agent were presented by *L. reuteri*. Therefore, *L. helveticus* or *L. fermentum* can be suggested as good alternatives to carry out the fermentation of BCJ. This is mainly because *L. fermentum* and *L. helveticus* showed higher CE activity (section 5.3) without the generation of a bacteriocin. After fermentation, the color of the different samples was analyzed and compared with the non-fermented juice at the initial time and after 31 d of fermentation (table 5.8). Results show that temperature and pH are two of the most important factors affecting anthocyanin stability and, therefore, their color. Greater differences were found between the non-fermented juice used as control at the initial time and after 31 d of analysis. Black carrot juice showed a higher lightness ($L^*$) value before fermentation. This value can be associated with the concentration of anthocyanins. As result of anthocyanins degradation and precipitation, dark products can be presented (Sarni-Manchado et al., 1996), this could be confirmed after 31 d with an increase in $L^*$ value rendering a darker juice.

**Table 5.8.** CIELab coordinates L*, a*, b*, C*, and h° color values of black carrot juice before and after fermentation [a]

| Sample | L* | a* | b* | C* | h° |
|---|---|---|---|---|---|
| BCJ (T= 0) | 17.31 ± 0.85 | 1.23 ± 0.15 | 2.43 ± 0.00 | 2.73 ± 0.78 | 64.25 ± 2.89 |
| BCJ (T= 31 days) | 16.69 ± 0.10 | 2.38 ± 0.07 | 2.17 ± 0.02 | 3.23 ± 0.04 | 42.36 ± 1.07 |
| *L. reuteri / L. mali* | 16.49 ± 0.01 | 1.77 ± 0.03 | 2.09 ± 0.04 | 2.74 ± 0.03 | 49.79 ± 0.73 |
| *L. reuteri / L. plantarum* | 16.46 ± 0.01 | 1.87 ± 0.05 | 2.05 ± 0.01 | 2.77 ± 0.04 | 47.73 ± 0.74 |
| *L. reuteri / L. sakei* | 16.51 ± 0.01 | 1.87 ± 0.07 | 2.07 ± 0.03 | 2.79 ± 0.02 | 47.96 ± 1.63 |
| BCJ (section 5.5.2) | 17.47 ± 0.08 | 3.15 ± 0.05 | 2.64 ± 0.06 | 4.11 ± 0.03 | 40.00 ± 1.05 |

[a] Values are mean ± standard deviation (n = 3)

An increase in L* value could also be observed by fermented juices together with an increase in the redness (a*) value in comparison with the initial juice which was observed due to pyranoanthocyanins content. These results are in agreement with those reported by (Sarni-Machado et al., 1996; Romero & Bakker, 2000). As reported by Rein and Heinonen (2004), an increase in chroma (C*) is expected by addition of copigments. Due to the formation of pyranoanthocyanins, all samples had an increase in C*. The higher C* observed in non-fermented juice after 31 d can be related to the decrease in L* value. As expected, calculation of hue angle value showed a decrease after 31 d in non-fermented juice used as control to assess anthocyanin degradation and decrease in the pH-value. Likewise, the color intensity was related to pyranoanthocyanins concentration, since hue angle calculation showed orange tonalities found in fermented juice and increased with an increase in the pyranoanthocyanins concentration (figure 5.16). As it is possible to observe in figure 5.16, samples of fermented juice (1-3), presented a more orange color characteristic compared to the original BCJ before (4) or after fermentation (5).

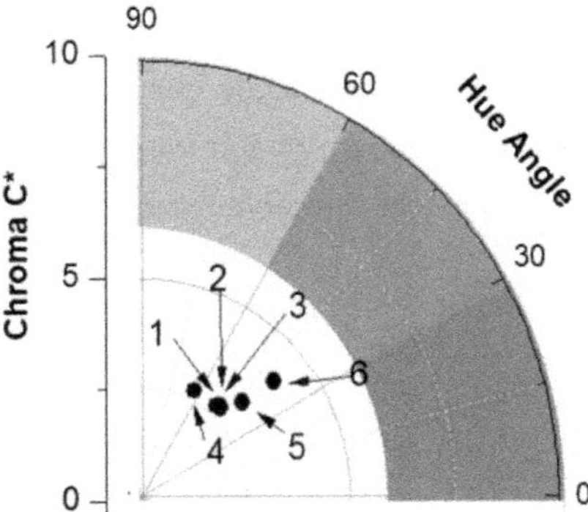


**Figure 5.16.** Color metric of black carrot juice (LCH color space). Black carrot juice fermented by *L. reuteri* and *L. mali* (1), *L. plantarum* (2), or *L. sakei* (3), non-fermented juice after 31 d (4), black carrot juice at T=0 (5), and black carrot juice fermented in section 5.5.2 (6)

The total color difference ($\Delta E^*$) was higher in the non-fermented juice used as control in comparison with the treatments containing pyranoanthocyanins (Table 5.9).

**Table 5.9.** The total color difference ($\Delta E^*$) between fermented juices and black carrot juice used as control before and after fermentation

| Sample | ΔE |
|---|---|
| BCJ (T= 0) / *L. reuteri* / *L. mali* | 1.11 |
| BCJ (T= 0) / *L. reuteri* / *L. plantarum* | 1.20 |
| BCJ (T= 0) / *L. reuteri* / *L. sakei* | 1.15 |
| BCJ (T= 31 days) | 1.38 |
| BCJ (T= 31) / *L. reuteri* / *L. mali* | 0.65 |
| BCJ (T= 31) / *L. reuteri* / *L. plantarum* | 0.57 |
| BCJ (T= 31) / *L. reuteri* / *L. sakei* | 0.55 |
| BCJ (T= 0) / BCJ (section 5.5.2) | 1.93 |

It has been demonstrated in previous sections (5.5.1) that pyranoanthocyanins possess higher stability in a greater pH range compared to their anthocyanins counterparts and present a higher stability to temperature exposure and color intensity. Therefore, the main advantage presented by the fermented juice is the color stability, since many treatments in the food industry involve temperature exposure where many anthocyanins can be degraded.

The results shown in this study demonstrate a very attractive approach to produce highly stable pigments like pyranoanthocyanins by the formulation of a coloring foodstuff as an alternative for the food industry.

# 6. CONCLUSION

The color presented by food directly affects taste and quality expectations. For this reason, food additives are commonly used in the food industry to enhance the acceptance of a product. Therefore, the development of this coloring foodstuff based on fermented black carrot juice represents a good alternative for substituting synthetic colorants used currently by the food industry.

Results showed that all LAB strains possessing cinnamoyl decarboxylase activity or esterase cinnamoyl activity carried out enzymatic activities with high yields of the corresponding products. These enzymatic products participate in the synthesis of pyranoanthocyanins. Cinnamoyl decarboxylase activity were assessed for *L. plantarum*, *L. mali*, and *L. sakei*. High decarboxylation rates (100%) were found when using *p*-CA and CA as substrates by these strains.

Additionally, *L. fermentum*, *L. helveticus*, and *L. reuteri* were investigated regarding their CE activity. Results showed that these strains displayed high hydrolysis rates of 5-CQA into caffeic acid. Furthermore, investigation of the metabolism of monochlorogenic and dichlorogenic acids into caffeic acid showed different specificities by these LAB strains. A novel pathway was revealed for the metabolism of 4-CQA comprising isomerase accompanied by an esterase activity by *L. helveticus*, and *L. reuteri*.

Synthesis of pyranoanthocyanins by fermentation of black carrot juice showed a strain-dependent variability in the metabolism of CA. While *L. reuteri* was shown to be able to hydrolyze 5-CQA and carry out decarboxylase activity of CA, the possible presence of a bacteriocin produced by *L. reuteri* could probably be responsible or the growth inhibition of *L. plantarum*, *L. mali*, and *L. sakei*. This result directly affects the production of 4-VC. However, high conversion rates of anthocyanins into pyranoanthocyanins were found (47%).

Black carrot juice and black carrot anthocyanins are well-known sources used for food coloring. The results of this study revealed the possibility of the use of LAB possessing cinnamoyl esterase activity and cinnamoyl decarboxylase activity for the generation of a coloring foodstuff rich in pyranoanthocyanins by fermentation of black carrot juice. A formulation of a coloring foodstuff represents several advantages like avoiding the use of artificial colorants and potential health benefits. This novel approach to synthesize pyranoanthocyanins by LAB fermentation was possible due to high concentrations in phenolic acids and anthocyanins of BCJ, which are precursors of the highly stable pyranoanthocyanins. Recent improvements suggest that biotechnology plays an important role in the food coloring field. In the food industry, Lactic acid bacteria are the most common started cultures. The use of Lactic acid bacteria and black carrot juice to produce a coloring foodstuff rich in pyranoanthocyanins offers an eco-friendly alternative to be applied around the world. Further protocols should be established to order of stabilize the pigments and to produce an applicable powder.

# 7. REFERENCES

Abeijón Mukdsi, M. C., Gauffin Cano, M. P., González, S. N., & Medina, R. B. (2012). Administration of *Lactobacillus fermentum* CRL1446 increases intestinal feruloyl esterase activity in mice. *Letters in Applied Microbiology, 54(1),* 18-25.

Aberoumand, A. (2011). A review article on edible pigments properties and sources as natural biocolorants in foodstuff and food industry. *World Journal of Dairy and Food Sciences, 6(1),* 71–78.

Adams, J. B. (1973). Thermal degradation of anthocyanins with particular reference to the 3-glycosides of cyanidin. I. In acidified aqueous solution at 100 °C. *Journal of the Science of Food and Agriculture*, *24(7)*, 747-762.

Aertsen, A., & Michiels, C. W. (2004). Stress and how bacteria cope with death and survival. *Critical Reviews in Microbiology*, *30(4)*, 263–273.

Ahmed, J., Shivhare, U. S., & Raghavan, G. S. V. (2004). Thermal degradation kinetics of anthocyanin and visual colour of plum puree. *European Food Research and Technology*, *218(6)*, 525–528.

Akihisa, T., Tokuda, H., Yasukawa, K., Ukiya, M., Kiyota, A., Sakamoto, N., Suzuki, T., Tanabe, N., & Nishino, H. (2005). Azaphilones, furanoisophthalides, and amino acids from the extracts of *Monascus pilosus*-fermented rice (red-mold rice) and their chemopreventive effects. *Journal of Agricultural and Food Chemistry*, *53(3)*, 562–565.

Alcalde-Eon, C., Escribano-Bailón, M.T., Santos-Buelga, C., & Rivas-Gonzalo, J.C. (2004). Separation of pyranoanthocyanins from red wine by column chromatography. *Analytica Chimica Acta*, *513*, 305–318.

Ali, K., Maltese, F., Choi, Y. H., & Verpoorte, R. (2010). Metabolic constituents of grapevine and grape-derived products. *Phytochemistry Reviews, 9 (3),* 357–378.

Alihosseini, F., Ju, K. S., Lango, J., Hammock, B. D., & Sun, G. (2008). Antibacterial colorants: characterization of prodiginines and their applications on textile materials. *Biotechnology Progress, 24 (3)*, 742–747.

An, G.H., Schuman, D. B., & Johnson, E. A. (1989). Isolation of *Phaffia rhodozyma* mutants with increased astaxanthin content. *Journal: Applied and Environmental Microbiology, 55(1),* 116–124.

Arscott, S. A., & Tanumihardjo, S. A. (2010). Carrots of many colors provide basic nutrition and bioavailable phytochemicals acting as a functional food. *Comprehensive Reviews in Food Science and Food Safety*, *9 (2)*, 223–239.

Asen, S., Stewart, R. N., & Norris, K. H. (1972). Co-pigmentation of anthocyanins in plant tissues and its effect on color. *Phytochemistry, 11(3),* 1139–1144.

Asenstorfer, R. E., Hayasaka, Y., & Jones, G. P. (2001). Isolation and structures of oligomeric wine pigments by bisulfite-mediated ion-exchange chromatography. *Journal of Agricultural and Food Chemistry, 49(12),* 5957–5963.

Attoe, E. L., & Elbe, J. H. (1981). Photochemial degradation of betanine and selected anthocyanins. *Journal of Food Science, 46(6),* 1934–1937.

Bai, Y., Findlay, B., Sanchez Maldonado, A. F., Schieber, A., Vederas, J. C., & Gänzle, M. G. (2014). Novel pyrano and vinylphenol adducts of deoxyanthocyanidins in sorghum sourdough. *Journal of Agricultural and Food Chemistry, 62(47),* 11536–11546.

Bakker, J., & Timberlake, C. F. (1997). Isolation, identification, and characterization of new color-stable anthocyanins occurring in some red wines. *Journal of Agricultural and Food Chemistry*, *45(1)*, 35-43.

Bąkowska-Barczak, A. (2005). Acylated anthocyanins as stable, natural food colorants - a review. *Polish Journal of Food and Nutrition Sciences, 14/55(2),*107-116.

Barthelmebs, L., Divies, C., & Cavin, J. F. (2000). Knockout of the *p*-coumarate decarboxylase gene from *Lactobacillus plantarum* reveals the existence of two other inducible enzymatic activities involved in phenolic acid metabolism. *Applied and Environmental Microbiology, 66(8),* 3368–3375.

Basílio, N., & Pina, F. (2016). Chemistry and photochemistry of anthocyanins and related compounds: a thermodynamic and kinetic approach. *Molecules*, *21(11),* 1502.

Bateman, B., Warner, J. O., Hutchinson, E., Dean, T., Rowlandson, P., Gant, C., Grundy, J., Fitzgerald, C., & Stevenson, J. (2004). The effects of a double blind, placebo controlled, artificial food colourings and benzoate preservative challenge on hyperactivity in a general population sample of preschool children. *Archives of Disease in Childhood, 89(6),* 506–511.

Baublis, A., Spomer, A. R.T., & Berber-Jiménez, M. D. (1994). Anthocyanin pigments: comparison of extract stability. *Journal of Food Science*, *59(6)*, 1219–1221.

Bechtold, T., & Mussak, R. (2009). *Handbook of natural colorants.* Wiley Series in Renewable Resource. Chichester: Wiley.

Bel-Rhlid, R., Thapa, D., Kraehenbuehl, K., Hansen, C. E., & Fischer, L. (2013). Biotransformation of caffeoyl quinic acids from green coffee extracts by *Lactobacillus johnsonii* NCC 533. *AMB Express*, *3*, 28-34.

Benito, S., Palomero, F., Morata, A., Uthurry, C., & Suárez-Lepe, J. A. (2009). Minimization of ethylphenol precursors in red wines via the formation of pyranoanthocyanins by selected yeasts. *International Journal of Food Microbiology*, *132(2-3)*, 145–152.

Bishayee, A., Mbimba, T., Thoppil, R. J., Háznagy-Radnai, E., Sipos, P., Darvesh, A. S., Folkesson, H. G., & Hohmann, J. (2011). Anthocyanin-rich black currant (*Ribes nigrum* L.) extract affords chemoprevention against diethylnitrosamine-induced hepatocellular carcinogenesis in rats. *The Journal of Nutritional Biochemistry*, *22(11)*, 1035–1046.

Bishayee, A., Thoppil, R. J., Mandal, A., Darvesh, A. S., Ohanyan, V., Meszaros, J. G., Háznagy-Radnai, E., Hohmann, J., & Bhatia, D. (2011). Black currant phytoconstituents exert chemoprevention of diethylnitrosamine-initiated hepatocarcinogenesis by suppression of the inflammatory response. *Molecular Carcinogenesis, 52(4),* 304–317.

Bobe, I. U., & Michel, M. (2011). Processing and product design for natural food products. In *Proceedings of the 11th International Congress on Engineering and Food,* 687-688 Athens, Greece: International Congress on Engineering and Food, available at http://www.icef11.org/content/papers/fpe/FPE1272.pdf

Bordignon-Luiz, M. T., Gauche, C., Gris, E. F., & Falcão, L. D. (2007). Colour stability of anthocyanins from Isabel grapes (*Vitis labrusca* L.) in model systems. *LWT - Food Science and Technology, 40(4),* 594–599.

Borges, M. E., Tejera, R. L., Díaz, L., Esparza, P., & Ibáñez, E. (2012). Natural dyes extraction from cochineal (*Dactylopius coccus*). New extraction methods. *Food Chemistry, 132(4),* 1855–1860.

Boulton, R. (2001). The Copigmentation of anthocyanins and Its role in the color of red wine: a critical Review. *American Journal of Enology and Viticulture, 52(2),* 67–87.

Brod, F. C. A., Vernal, J., Borges Bertoldo, J., Terenzi, H., & Maissonave Arisi, A. C. (2010). Cloning, expression, purification, and characterization of a novel esterase from *Lactobacillus plantarum*. *Molecular Biotechnology, 44(3),* 242-249.

Brouillard, R. (1983). The in vivo expression of anthocyanin colour in plants. *Phytochemistry*, *22(6)*, 1311–1323.

Brouillard, R. (1988). Flavonoids and flower color. In *The flavonoids Advances in Research Since 1980*, 525–538.

Brouillard, R., & Dangles, O. (1994). Anthocyanin molecular interactions: the first step in the formation of new pigments during wine aging?. *Food Chemistry*, *51(4)*, 365–371.

Buron, N., Coton, M., Desmarais, C., Ledauphin, J., Guichard, H., Barillier, D., & Coton, E. (2011). Screening of representative cider yeasts and bacteria for volatile phenol-production ability. *Food Microbiology*, *28(7)*, 1243–1251.

Buttery, R. G., & Ling, L. C. (1995). Volatile flavor components of corn tortillas and related products. *Journal of Agricultural and Food Chemistry, 43(7)*, 1878–1882.

Buttery, R. G., Orts, W. J., Takeoka, G. R., & Nam, Y. (1999). Volatile flavor components of rice cakes. *Journal of Agricultural and Food Chemistry, 47(10)*, 4353–4356.

Cabrita, M. J., Freitas, A. M. C., Laureano, O., & Di Stefano, R. (2006). Glycosidic aroma compounds of some Portuguese grape cultivars. *Journal of the Science of Food and Agriculture, 86(6)*, 922–931.

Cabrita, M. J., Palma, V., Patão, R., & Freitas, A. M. C. (2012). Conversion of hydroxycinnamic acids into volatile phenols in a synthetic medium and in red wine by *Dekkera bruxellensis*. *Journal of Agricultural and Food Chemistry, 32(1)*, 106–112.

Cameira-dos-Santos, P.J., Brillouet, J.M., Cheynier, V., & Moutounet, M. (1996). Detection and partial characterisation of new anthocyanin-derived pigments in wine. *Journal of the Science of Food and Agriculture, 70(2)*, 204–208.

Campos, F. M., Couto, J. A., & Hogg, T. A. (2003). Influence of phenolic acids on growth and inactivation of *Oenococcus oeni* and *Lactobacillus hilgardii*. *Journal of Applied Microbiology, 94(2)*, 167-174.

Campos, F. M., Couto, J. A., Figueiredo, A. R., Tóth, I. V., Rangel, A. O. S. S., & Hogg, T. A. (2009). Cell membrane damage induced by phenolic acids on wine lactic acid bacteria. *Journal of Food Microbiology, 135(2)*, 144–151.

Cañamares, M. V., & Lombardi, J. R. (2015). Raman, SERS, and DFT of mauve dye: adsorption on Ag nanoparticles. *The Journal of Physical Chemistry C. 119(25)*, 142974–14303.

Carmelo, V., Santos, H., & Sá-Correia, I. (1997). Effect of extracellular acidification on the activity of plasma membrane ATPase and on the cytosolic and vacuolar pH of *Saccharomyces cerevisiae*. *Biochimica et Biophysica Acta (BBA) - Biomembranes, 1325(1)*, 63–70.

Castañeda-Ovando, A., Pacheco-Hernández, M. d. L., Páez-Hernández, M. E., Rodríguez, J. A., & Galán-Vidal, C. A. (2009). Chemical studies of anthocyanins: a review. *Food Chemistry, 113(4),* 859–871.

Castro, H. P., Teixeira, P. M., & Kirby, R. (1996). Changes in the cell membrane of *Lactobacillus bulgaricus* during storage following freeze-drying. *Biotechnology Letters, 18(1),* 99–104.

Cavin, J. F., Andioc, V., Etievant, P. X., & Divies, C. (1993). Ability of wine lactic acid bacteria to metabolize phenol carboxylic acids. *American Journal of Enology and Viticulture*, *44(1)*, 76–80.

Cavin, J. F., Barthelmebs, L., & Diviès, C. (1997). Molecular characterization of an inducible *p*-coumaric acid decarboxylase from *Lactobacillus plantarum*: gene cloning, transcriptional analysis, overexpression in *Escherichia coli*, purification, and characterization. *Applied Environmental Microbiology, 63(5),* 1939-1944.

Chatonnet, P., Dubourdie, D., Boidron, J. N, & Pons, M. (1992). The origin of ethylphenols in wines. *Journal of the Science of Food and Agriculture, 60(2),* 165–178.

Chatonnet, P., Dubourdieu, D., Boidron, J. N., & Lavigne, V. (1993). Synthesis of volatile phenols by *Saccharomyces cerevisiae* in wines. *Journal of the Science of Food and Agriculture*, *62(2)*, 191–202.

Chatonnet, P., Dubourdieu, D., & Boidron, J. N. (1995). The influence of *Brettanomyces/Dekkera* sp. yeasts and lactic acid bacteria on the ethylphenol content of red wines. *American Journal of Enology and Viticulture*, *46(4)*, 463–468.

Chatonnet, P., Viala, C., & Dubourdieu, D. (1997). Influence of polyphenolic components of red wines on the microbial synthesis of volatile phenols. *American Journal of Enology and Viticulture, 48(4),* 443-448.

Chattopadhyay, P., Chatterjee, S., & Sen, S. K. (2008). Biotechnological potential of natural food grade biocolorants. *African Journal of Biotechnology, 7(17)*, 2972-2985.

Chen, F., Long, X., Liu, Z., Shao, H., & Liu, L. (2014). Analysis of phenolic acids of Jerusalem artichoke (*Helianthus tuberosus* L.) responding to salt-stress by liquid chromatography/tandem mass spectrometry. *Scientific World Journal, 2014.* Article ID 568043. doi:10.1155/2014/568043

Cisse, M., Vaillant, F., Acosta, O., Dhuique-Mayer, C., & Dornier, M. (2009). Thermal degradation kinetics of anthocyanins from blood orange, blackberry, and roselle using the arrhenius, eyring, and ball models. *Journal of Agricultural and Food Chemistry, 57(14),* 6285–6291.

Clifford, M. N. (1999). Chlorogenic acids and other cinnamates – nature, occurrence and dietary burden. *Journal of the Science of Food and Agriculture, 79(3),* 362–372.

Clifford, M. N. (2000a). Anthocyanins - nature, occurrence and dietary burden. *Journal of the Science of Food and Agriculture*, *80(7)*, 1063–1072.

Clifford, M. N. (2000b). Chlorogenic acids and other cinnamates - nature, occurrence, dietary burden, absorption and metabolism. *Journal of the Science of Food and Agriculture*, *80(7)*, 1033–1043.

Clifford, M. N., Knight, S., & Kuhnert, N. (2005). Discriminating between the six isomers of dicaffeoylquinic acid by LC-MS$^n$. *Journal of Agricultural and Food Chemistry*, *53(10)*, 3821–3832.

Clifford, M. N., Knight, S., Surucu, B., & Kuhnert, N. (2006). Characterization by LC-MS$^n$ of four new classes of chlorogenic acids in green coffee beans: dimethoxycinnamoylquinic acids, diferuloylquinic acids, caffeoyl-dimethoxycinnamoylquinic acids, and feruloyl-dimethoxycinnamoylquinic acids. *Journal of Agricultural and Food Chemistry*, *54(6)*, 1957–1969.

Clydesdale, F. M. (1991). Color perception and food quality. *Journal of Food Quality*, *14(1)*, 61–74.

Combes, R. D., & Haveland-Smith, R. B. (1982). A review of the genotoxicity of food, drug and cosmetic colours and other azo, triphenylmethane and xanthene dyes. *Mutation Research/Reviews in Genetic Toxicology*, *98(2)*, 101–243.

Couteau, D., McCartney, A. L., Gibson, G. R., Williamson, G., & Faulds, C. B. (2001). Isolation and characterization of human colonic bacteria able to hydrolyse chlorogenic acid. *Journal of Applied Microbiology, 90(6),* 873-881.

Couto, J. A., Campos, F. M., Figueiredo, A. R., & Hogg, T. A. (2006). Ability of lactic acid bacteria to produce volatile phenols. *American Journal of Enology and Viticulture*, *57(2)*, 166–171.

Crepin, V. F., Faulds, C. B., & Connerton, I. F. (2004). Functional classification of the microbial feruloyl esterases. *Applied Microbiology and Biotechnology, 63(6),* 647-652.

Cruz, L., Brás, N. F., Teixeira, N., Mateus, N., Ramos, M. J., Dangles, O., & de Freitas, V. (2010). Vinylcatechin dimers are much better copigments for anthocyanins than catechin dimer procyanidin B3. *Journal of Agricultural and Food Chemistry, 58(5),* 3159–3166.

Curiel, J. A., Rodríguez, H., Landete, J. M., de las Rivas, B., & Muñoz, R. (2010). Ability of *Lactobacillus brevis* strains to degrade food phenolic acids. *Food Chemistry, 120(1),* 225–229.

Dabas, D., Elias, R. J., Lambert, J. D., & Ziegler, G. R. (2011). A colored avocado seed extract as a potential natural colorant. *Journal of Food Science, 76(9),* C1335-C1341.

Dao, L., & Friedman, M. (1992). Chlorogenic acid content of fresh and processed potatoes determined by ultraviolet spectrophotometry. *Journal of Agricultural and Food Chemistry*, *40(11)*, 2152–2156.

de Freitas, V., & Mateus, N. (2011). Formation of pyranoanthocyanins in red wines: a new and diverse class of anthocyanin derivatives. *Analytical and Bioanalytical Chemistry, 401(5),* 1463–1473.

de Las Rivas, B., Rodríguez, H., Curiel, J. A., Landete, J. M., & Muñoz, R. (2009). Molecular screening of wine lactic acid bacteria degrading hydroxycinnamic acids. *Journal of Agricultural and Food Chemistry, 57(2),* 490-494.

Degrassi, G., Polverino de Laureto, P., & Bruschi, C. V. (1995). Purification and characterization of ferulate and *p*-coumarate decarboxylase from

*Bacillus pumilus. Applied and Environmental Microbiology, 61(1),* 326–332.

Del Pozo-Insfran, D., Del Follo-Martinez, A., Talcott, S. T., & Brenes, C. H. (2007). Stability of copigmented anthocyanins and ascorbic acid in muscadine grape juice processed by high hydrostatic pressure. *Journal of Food Science, 72(4),* 247-253.

Delgado-Vargas, F., Jiménez, A. R., & Paredes-López, O. (2000). Natural pigments: carotenoids, anthocyanins, and betalains--characteristics, biosynthesis, processing, and stability. *Critical Reviews in Food Science and Nutrition, 40(3),* 173–289.

Delgado-Vargas, F., & Paredes-Lopez, O. (2002). *Natural colorants for food and nutraceutical uses.* Boca Raton, FL: CRC Press.

Deshpande, S., Jaiswal, R., Matei, M. F., & Kuhnert, N. (2014). Investigation of acyl migration in mono- and dicaffeoylquinic acids under aqueous basic, aqueous acidic, and dry roasting conditions. *Journal of Agricultural and Food Chemistry, 62 (37),* 9160-9170.

Dilokpimol, A., Mäkelä, M. R., Aguilar-Pontes, M. V., Benoit-Gelber, I., Hildén, K. S., & de Vries, R. P. (2016). Diversity of fungal feruloyl esterases: updated phylogenetic classification, properties, and industrial applications. *Biotechnology for Biofuels, 9(1),* 231.

Downham, A., & Collins, P. (2000). Colouring our foods in the last and next millennium. *International Journal of Food Science and Technology, 35(1),* 5–22.

Dufossé, L. (2006). Microbial Production of Food Grade Pigments. *Food Technology and Biotechnology, 44(3),* 313-323.

Dufossé, L., Fouillaud, M., Caro, Y., Mapari, S. A. S., & Sutthiwong, N. (2014). Filamentous fungi are large-scale producers of pigments and colorants for the food industry. *Current Opinion in Biotechnology, 26,* 56–61.

Dugo, P., Mondello, L., Errante, G., Zappia, G., & Dugo, G. (2001). Identification of anthocyanins in berries by narrow-bore high-performance liquid chromatography with electrospray ionization detection. *Journal of Agricultural and Food Chemistry, 49(8),* 3987-3992.

Edlin, D. A.N., Narbad, A., Gasson, M. J., Dickinson, J. R., & Lloyd, D. (1998). Purification and characterization of hydroxycinnamate decarboxylase from *Brettanomyces anomalus. Enzyme and Microbial Technology, 22(4),* 232–239.

Eiro, M. J., & Heinonen, M. (2002). Anthocyanin color behavior and stability during storage: effect of intermolecular copigmentation. *Journal of Agricultural and Food Chemistry, 50(25),* 7461–7466.

Elbanna, K., Sarhan, O. M., Khider, M., Elmogy, M., Abulreesh, H. H., & Shaaban, M. R. (2017). Microbiological, histological, and biochemical evidence for the adverse effects of food azo dyes on rats. *Journal of Food and Drug Analysis, 25(3),* 667–680.

Esatbeyoglu, T., Rodríguez-Werner, M., Schlösser, A., Liehr, M., Ipharraguerre, I., Winterhalter, P., & Rimbach, G. (2016). Fractionation of plant bioactives from black carrots (*Daucus carota* subspecies *sativus* varietas *atrorubens* Alef.) by adsorptive membrane chromatography and analysis of their potential anti-diabetic activity. *Journal of Agricultural and Food Chemistry, 64(29),* 5901–5908.

Escribano-Bailón, M. T., & Santos-Buelga, C. (2012). Anthocyanin copigmentation-evaluation, mechanisms and implications for the colour of red wines. *Current Organic Chemistry, 16(6),* 715–723.

Esteban-Torres, M., Reverón, I., Mancheño, J. M., de las Rivas, B., & Muñoz, R. (2013). Characterization of a feruloyl esterase from *Lactobacillus plantarum. Applied Environmental Microbiology, 79(17*), 5130-5136.

Esteban-Torres, M., Landete, J. M., Reverón, I., Santamaría, L., de las Rivas, B., & Muñoz, R. (2015). A *Lactobacillus plantarum* esterase active on a broad range of phenolic esters. *Applied and Environmental Microbiology, 81(9),* 3235–3242.

Esteves Torres, F. A., Zaccarim, B. R., de Lencastre Novaes, L. C., Jozala, A. F., dos Santos, C. A., Simas Teixeira, M. F., & Santos-Ebinuma, V.

C. (2016). Natural colorants from filamentous fungi. *Applied Microbiology and Biotechnology, 100(6),* 2511–2521.

Fabre, C. E., Santerre, A. L., Loret, M. O., Baberian, R., Pareilleux, A., Goma, G., & Blanc, P. J. (1993). Production and food applications of the red pigments of *Monascus ruber*. *Journal of Food Science, 58(5),* 1099–1102.

Fan, G., Han, Y., Gu, Z., & Gu, F. (2008). Composition and colour stability of anthocyanins extracted from fermented purple sweet potato culture. *LWT - Food Science and Technology, 41(8)*, 1412–1416.

Fan-Chiang, H. J., & Wrolstad, R. E. (2005). Anthocyanin pigment composition of blackberries. *Journal of Food Science, 70(3),* C198-C202.

Felis, G. E., & Dellaglio, F. (2007). Taxonomy of Lactobacilli and Bifidobacteria. *Current Issues in Intestinal Microbiology*, *8(2)*, 44–61.

Fenster, K. M., Parkin, K. L., & Steele, J. L. (2003). Intracellular esterase from *Lactobacillus casei* LILA: Nucleotide sequencing, purification, and characterization. *Journal of Dairy Science, 86(4),* 1118-1129.

Fernández Murga, M. L., Cabrera, G. M., Front de Valdez, G., Disalvo, A., & Seldes, A. M. (2000). Influence of growth temperature on cryotolerance and lipid composition of *Lactobacillus acidophilus*. *Journal of Applied Microbiology, 88(2),* 342–348.

Fernández-López, J. A., Angosto, J. M., Giménez, P. J., & León, G. (2013). Thermal stability of selected natural red extracts used as food colorants. *Plant Foods for Human Nutrition, 68(1),* 11–17.

Fiddler, W., Parker, W. E., Wasserman, A. E., & Doerr, R. C. (1967). Thermal decomposition of ferulic acid. *Journal of Agricultural and Food Chemistry, 15(5),* 757–761.

Figueiredo, P., Elhabiri, M., Toki, K., Saito, N., Dangles, O., & Brouillard, R. (1996). New aspects of anthocyanin complexation. Intramolecular copigmentation as a means for colour loss?. *Phytochemistry, 41(1),* 301–308.

Filannino, P., Gobbetti, M., De Angelis, M., & Di Cagno, R. (2014). Hydroxycinnamic acids used as external acceptors of electrons: an energetic advantage for strictly heterofermentative lactic acid bacteria. *Applied Environmental Microbiology, 80(24),* 7574-7582.

Filannino, P., Bai, Y., Di Cagno, R., Gobbetti, M., & Gänzle, M. G. (2015). Metabolism of phenolic compounds by *Lactobacillus* spp. during fermentation of cherry juice and broccoli puree. *Food Microbiology, 46,* 272–279.

Filannino, P., Di Cagno, R., & Gobbetti, M. (2018). Metabolic and functional paths of lactic acid bacteria in plant foods: get out of the labyrinth. *Current Opinion in Biotechnology, 49,* 64–72.

Frick, D. (2003). The coloration of food. *Review of Progress in Coloration and Related Topics, 33(1),* 15–32.

Fritsch, C., Jänsch, A., Ehrmann, M. A., Toelstede, S., & Vogel, R. F. (2017). Characterization of cinnamoyl esterases from different Lactobacilli and Bifidobacteria. *Current Microbiology, 74(2),* 247–256.

Fulcrand, H., Cameira dos Santos, P.J., Sarni-Manchado, P., Cheynier, V., & Favre-Bonvin, J. (1996). Structure of new anthocyanin-derived wine pigments. *Journal of the Chemical Society, Perkin Transactions 1(7),* 735.

Fulcrand, H., Benabdeljalil, C., Rigaud, J., Cheynier, V., & Moutounet, M. (1998). A new class of wine pigments generated by reaction between pyruvic acid and grape anthocyanins. *Phytochemistry, 47(7),* 1401–1407.

Gänzle, M. G., Höltzel, A., Walter, J., Jung, G., & Hammes, W. P. (2000). Characterization of reutericyclin produced by *Lactobacillus reuteri* LTH2584. *Applied Environmental Microbiology, 66(10),* 4325-4333.

Garcia-Alonso, M., Rimbach, G., Sasai, M., Nakahara, M., Matsugo, S., Uchida, Y., Rivas-Gonzalo, J. C., & De Pascual-Teresa, S. (2005). Electron spin resonance spectroscopy studies on the free radical scavenging activity of wine anthocyanins and pyranoanthocyanins. *Molecular Nutrition and Food Research, 49(12),* 1112-1119.

García-Viguera, C., Zafrilla, P., Romero, F., Abellán, P., Artés, F., & Tomás-Barberán, F. A. (1999). Color stability of strawberry jam as affected by cultivar and storage temperature. *Journal of Food Science, 64(2),* 243-247.

Giusti, M.M., & Wrolstad, R. E. (2003). Acylated anthocyanins from edible sources and their applications in food systems. *Biochemical Engineering Journal, 14(3),* 217–225.

Gómez Gallego, M. A., Gómez García-Carpintero, E., Sánchez-Palomo, E., González Viñas, M. A., & Hermosín-Gutiérrez, I. (2013). Evolution of the phenolic content, chromatic characteristics and sensory properties during bottle storage of red single-cultivar wines from Castilla La Mancha region. *Food Research International, 51(2),* 554–563.

Gonthier, M. P., Verny, M. A., Besson, C., Rémésy, C., & Scalbert, A. (2003). Chlorogenic acid bioavailability largely depends on its metabolism by the gut microflora in rats. *The Journal of Nutrition, 133(6),* 1853-1859.

González-Manzano, S., Santos-Buelga, C., Dueñas, M., Rivas-Gonzalo, J. C., & Escribano-Bailón, T. (2008). Colour implications of self-association processes of wine anthocyanins. *European Food Research and Technology, 226(3),* 483–490.

Gosset, G. (2009). Production of aromatic compounds in bacteria. *Current Opinion in Biotechnology, 20(6),* 651-658.

Goto, T., & Kondo, T. (1991). Structure and molecular stacking of anthocyanins—flower color variation. *Angewandte Chemie International Edition in English*, *30(1)*, 17–33.

Gras, C. C., Bogner, H., Carle, R., & Schweiggert, R. M. (2016). Effect of genuine non-anthocyanin phenolics and chlorogenic acid on color and stability of black carrot (*Daucus carota* ssp. *sativus* var. *atrorubens* Alef.) anthocyanins. *Food Research International*, *85*, 291–300.

Groot, M. N. N., & de Bont, J. A. (1998). Conversion of phenylalanine to benzaldehyde initiated by an aminotransferase in *Lactobacillus plantarum*. *Applied Environmental Microbiology, 64(8),* 3009-3013.

Guerzoni, M. E., Lanciotti, R., & Cocconcelli, P. S. (2001). Alteration in cellular fatty acid composition as a response to salt, acid, oxidative and thermal stresses in *Lactobacillus helveticus. Microbiology, 147(8),* 2255–2264.

Guglielmetti, S., de Noni, I., Caracciolo, F., Molinari, F., Parini, C., & Mora, D. (2008). Bacterial cinnamoyl esterase activity screening for the production of a novel functional food product. *Applied and Environmental Microbiology, 74(4),* 1284–1288.

Harasym, J., & Bogacz-Radomska, L. (2016). Colorants in foods - from past to present. *Engineering Science and Technologies / Nauki Inżynierskie i Technologie, 3(22),* 21–35.

Hayasaka, Y., & Asenstorfer, R. E. (2002). Screening for potential pigments derived from anthocyanins in red wine using nano-electrospray tandem mass spectrometry. *Journal of Agricultural and Food Chemistry, 50(4),* 756–761.

Hayashi, K., Ohara, N., & Tsukui, A. (1996). Stability of anthocyanins in various vegetables and fruits. *Food Science and Technology International, 2(1),* 30–33.

He, J., Santos-Buelga, C., Silva, A. M. S., Mateus, N., & de Freitas, V. (2006). Isolation and structural characterization of new anthocyanin-derived yellow pigments in aged red wines. *Journal of Agricultural and Food Chemistry, 54(25),* 9598–9603.

He, J., Carvalho, A. R. F., Mateus, N., & de Freitas, V. (2010a). Spectral features and stability of oligomeric pyranoanthocyanin-flavanol pigments isolated from red wines. *Journal of Agricultural and Food Chemistry, 58(16),* 9249–9258.

He, J., Oliveira, J., Silva, A. M. S., Mateus, N., & de Freitas, V. (2010b). Oxovitisins: a new class of neutral pyranone-anthocyanin derivatives in red wines. *Journal of agricultural and Food Chemistry, 58(15),* 8814–8819.

He, J. J., Han, F. L., Yu, Q. Q., Pan, Q. H., Duan, C. Q., & Cheng, G. L. (2010c). Effects of the maceration enzymes on evolution of pyranoanthocyanins and cinnamic acids during the cabernet gernischet (*Vitis vinifera* L. cv.) red wine making. *Food Science and Biotechnology, 19(3),* 603-610.

Heer, K., & Sharma, S. (2017). Microbial pigments as a natural color: a review. *International Journal of Pharmaceutical Sciences and Research, 8(5),*1913-1922.

Hendry, G. A. F., & Houghton, J. D. (1996). *Natural food colorants.* 2nd ed. Blackie and Professional. London, UK.

Herbert, P., Cabrita, M. J., Ratola, N., Laureano, O., & Alves, A. (2005). Free amino acids and biogenic amines in wines and musts from the Alentejo region. Evolution of amines during alcoholic fermentation and relationship with variety, sub-region and vintage. *Journal of Food Engineering, 66(3),* 315–322.

Hernández-Almanza, A., Montanez, J.C., Aguilar-González, M. A., Martínez-Ávila, C., Rodríguez-Herrera, R., & Aguilar, C. N. (2014). *Rhodotorula glutinis* as source of pigments and metabolites for food industry. *Food Bioscience*, *5*, 64–72.

Herrmann, K., & Nagel, C. W. (1989). Occurrence and content of hydroxycinnamic and hydroxybenzoic acid compounds in foods. *Critical Reviews in Food Science and Nutrition*, *28(4)*, 315–347.

Hillebrand, S., Schwarz, M., & Winterhalter, P. (2004). Characterization of anthocyanins and pyranoanthocyanins from blood orange [*Citrus sinensis* (L.) Osbeck] juice. *Journal of Agricultural and Food Chemistry*, *52(24)*, 7331–7338.

Hisano, A. (2016a). The rise of synthetic colors in the american food industry, 1870–1940. *Business History Review*, *90*(3), 483-504.

Hisano, A. (2016b). Standardized color in the food industry: the co-creation of the food coloring business in the United States, 1870-1940. *Harvard Business School General Management Unit Working Paper*, 17-037.

Hole, A. S., Rud, I., Grimmer, S., Sigl, S., Narvhus, J., & Sahlstrøm, S. (2012). Improved bioavailability of dietary phenolic acids in whole grain barley and oat groat following fermentation with probiotic *Lactobacillus acidophilus*, *Lactobacillus johnsonii*, and *Lactobacillus reuteri*. *Journal of Agricultural and Food Chemistry, 60(25),* 6369-6375.

Horbowicz, M., Kosson, R., Grzesiuk, A., & Dębski, H. (2008). Anthocyanins of fruits and vegetables-their occurrence, analysis and role in human nutrition. *Vegetable Crops Research Bulletin*, *68*, 5-22.

Hua, D. L., Liang, X. H., Zhang, X. D., Zhang, J., Xu, H. P., Li, Y., & Xu, P. (2013). Rapid Transformation of Ferulic Acid to 4-Vinyl Guaiacol by *Bacillus pumilus* S-1. *Asian Journal of Chemistry, 25(2), 615–617.*

Huang, Z., Dostal, L., & Rosazza, J. P. (1994). Purification and characterization of a ferulic acid decarboxylase from *Pseudomonas fluorescens*. *Journal of Bacteriology, 176(19),* 5912–5918.

Hutkins, R. W., & Nannen, N. L. (1993). pH homeostasis in lactic acid bacteria. *Journal of Dairy Science, 76(8),* 2354–2365.

Ignat, I., Volf, I., & Popa, V. I. (2011). A critical review of methods for characterisation of polyphenolic compounds in fruits and vegetables. *Food Chemistry, 126(4),* 1821–1835.

Jaiswal, V., DerMarderosian, A., & Porter, J. R. (2010). Anthocyanins and polyphenol oxidase from dried arils of pomegranate (*Punica granatum* L.). *Food Chemistry*, *118(1)*, 11–16.

Jiang, D., & Peterson, D. G. (2010). Role of hydroxycinnamic acids in food flavor: a brief overview. *Phytochemistry Reviews*, *9(1)*, 187–193.

Johnson, E. A., & Lewis, M. J. (1979). Astaxanthin formation by the yeast *Phaffia rhodozyma*. *Journal of General Microbiology*, *115(1)*, 173–183.

Joshi, V. K., Attri, D., Bala, A., & Bhushan, S. (2003). Microbial Pigments. *Indian Journal of Biotechnology*, *2*, 362–369.

Jung, D. H., Choi, W., Choi, K. Y., Jung, E., Yun, H., Kazlauskas, R. J., & Kim, B. G. (2013). Bioconversion of *p*-coumaric acid to *p*-hydroxystyrene using phenolic acid decarboxylase from *B. amyloliquefaciens* in biphasic reaction system. *Applied Microbiology and Biotechnology, 97(4),* 1501-1511.

Kabera, J. N., Semana, E., Mussa, A. R., & He, X. (2014). Plant secondary metabolites: biosynthesis, classification, function and pharmacological properties. *Journal of Pharmacy and Pharmacology, 2*, 377-392.

Kammerer, D., Carle, R., & Schieber, A. (2003). Detection of peonidin and pelargonidin glycosides in black carrots (*Daucus carota* ssp. *sativus* var. *atrorubens* Alef.) by high-performance liquid chromatography/electrospray ionization mass spectrometry. *Rapid Communications in Mass Spectrometry*, *17(21)*, 2407–2412.

Kammerer, D., Carle, R., & Schieber, A. (2004a). Characterization of phenolic acids in black carrots (*Daucus carota* ssp. *sativus* var. *atrorubens* Alef.) by high-performance liquid chromatography/electrospray ionization mass spectrometry. *Rapid Communications in Mass Spectrometry, 18(12),* 1331–1340.

Kammerer, D., Carle, R., & Schieber, A. (2004b). Quantification of anthocyanins in black carrot extracts (*Daucus carota* ssp. *sativus* var. *atrorubens* Alef.) and evaluation of their color properties. *European Food Research and Technology*, *219(5)*, 479–486.

Kammerer, D. R., Schillmöller, S., Maier, O., Schieber, A., & Carle, R. (2007). Colour stability of canned strawberries using black carrot and elderberry juice concentrates as natural colourants. *European Food Research and Technology, 224(6),* 667-679.

Kang, S. Y., Choi, O., Lee, J. K., Hwang, B. Y., Uhm, T. B., & Hong, Y. S. (2012). Artificial biosynthesis of phenylpropanoic acids in a tyrosine overproducing *Escherichia coli* strain. *Microbial Cell Factories, 11(1),* 153.

Kapusta, I., Krok, E. S., Jamro, D. B., Cebulak, T., Kaszuba, J., & Salach, R. T. (2013). Identification and quantification of phenolic compounds from Jerusalem artichoke (*Helianthus tuberosus* L.) tubers. *Journal of Food, Agriculture & Environment*, *11(3&4)*, 601–606.

Kawai, Y., Ishii, Y., Uemura, K., Kitazawa, H., Saito, T., & Itoh, T. (2001). *Lactobacillus reuteri* LA6 and *Lactobacillus gasseri* LA39 isolated from faeces of the same human infant produce identical cyclic bacteriocin. *Food Microbiology, 18(4),* 407-415.

Khandare, V., Walia, S., Singh, M., & Kaur, C. (2011). Black carrot (*Daucus carota* ssp. *sativus*) juice: processing effects on antioxidant composition and color. *Food and Bioproducts Processing, 89(4),* 482-486.

Khoo, H. E., Azlan, A., Tang, S. T., & Lim, S. M. (2017). Anthocyanidins and anthocyanins: colored pigments as food, pharmaceutical ingredients, and the potential health benefits. *Food and Nutrition Research, 61(1),* 1361779.

Kırca, A., Özkan, M., & Cemeroğlu, B. (2006). Stability of black carrot anthocyanins in various fruit juices and nectars. *Food Chemistry, 97(4),* 598–605.

Kırca, A., Özkan, M., & Cemeroğlu, B. (2007). Effects of temperature, solid content and pH on the stability of black carrot anthocyanins. *Food Chemistry, 101(1),* 212-218.

Kong, J.M., Chia, L.S., Goh, N.K., Chia, T.F., & Brouillard, R. (2003). Analysis and biological activities of anthocyanins. *Phytochemistry, 64(5),* 923–933.

Konings, W. N., Lolkema, J. S., & Poolman, B. (1995). The generation of metabolic energy by solute transport. *Archives of Microbiology*, *164(4)*, 235–242.

Konings, W. N. (2002). The cell membrane and the struggle for life of lactic acid bacteria. In Lactic acid bacteria. *Antonie van Leeuwenhoek, 82*, 3-27.

Kono, Y., Shibata, H., Kodama, Y., & Sawa, Y. (1995). The suppression of the N-nitrosating reaction by chlorogenic acid. *Biochemical Journal, 312(3)*, 947–953.

Kopjar, M., Bilić, B., & Piližota, V. (2011). Influence of different extracts addition on total phenols, anthocyanin content and antioxidant activity of blackberry juice during storage. *Croatian Journal of Food Science and Technology, 3(1),* 9-15.

Lai, K. K., Lorca, G. L., & Gonzalez, C. F. (2009). Biochemical properties of two cinnamoyl esterases purified from a *Lactobacillus johnsonii* strain isolated from stool samples of diabetes-resistant rats. *Applied Environmental Microbiology, 75(15),* 5018-5024.

Lai, K.K., Vu, C., Valladares, R. B., Potts, A., H. & Gonzalez, C., F. (2012). Identification and characterization of feruloyl esterases produced by probiotic bacteria, *In* Rizwan Ahmad (Ed.), *Protein Purification*, 151-166. InTech, Available from: https://www.intechopen.com/books/protein-purification/identification-and-characterization-of-feruloyl-esterases-produced-by-probiotic-bacteria

Lakshmi, C. G. (2014). Food coloring: the natural way. *Research Journal of Chemical Sciences, 4(2),* 87–96.

Landete, J. M., Ferrer, S., Polo, L., & Pardo, I. (2005). Biogenic amines in wines from three Spanish regions. *Journal of Agricultural and Food Chemistry, 53(4),* 1119–1124.

Larsson, S., Nilvebrant, N. O., & Jönsson, L. (2001). Effect of overexpression of *Saccharomyces cerevisiae* Pad1p on the resistance to phenylacrylic acids and lignocellulose hydrolysates under aerobic and oxygen-limited conditions. *Applied Microbiology and Biotechnology, 57(1-2),* 167-174.

Lee, M. H., Ho, C.T., & Chang, S. S. (1981). Thiazoles, oxazoles, and oxazolines identified in the volatile flavor of roasted peanuts. *Journal of Agricultural and Food Chemistry, 29(3)*, 684–686.

Lee, H. S., & Nagy, S. (1990). Formation of 4-vinyl guaiacol in adversely stored orange juice as measured by an improved HPLC method. *Journal of Food Science, 55(1),* 162-163.

Louie, G. V., Bowman, M. E., Moffitt, M. C., Baiga, T. J., Moore, B. S., & Noel, J. P. (2006). Structural determinants and modulation of substrate specificity in phenylalanine-tyrosine ammonia-lyases. *Chemistry and Biology, 13(12),* 1327-1338.

Lu, Y., & Foo, L.Y. (2001). Unusual anthocyanin reaction with acetone leading to pyranoanthocyanin formation. *Tetrahedron Letters, 42(7),* 1371–1373.

Lu, Y., Sun, Y., & Foo, L.Y. (2000). Novel pyranoanthocyanins from black currant seed. *Tetrahedron Letters, 41(31)*, 5975–5978.

Maccarone, E., Maccarrone, A., & Rapisarda, P. (1985). Stabilization of anthocyanins of blood orange fruit juice. *Journal of Food Science*, *50(4),* 901-904.

Madsen, H. L., Stapelfeldt, H., Bertelsen, G., & Skibsted, L. H. (1993). Cochineal as a colorant in processed pork meat. Colour matching and oxidative stability. *Food Chemistry*. *46*(3), 265-271.

Malien-Aubert, C., Dangles, O., & Amiot, M. J. (2001). Color stability of commercial anthocyanin-based extracts in relation to the phenolic composition. Protective effects by intra-and intermolecular copigmentation. *Journal of Agricultural and Food Chemistry*, *49(1),* 170–176.

Manach, C., Scalbert, A., Morand, C., Rémésy, C., & Jiménez, L. (2004). Polyphenols: food sources and bioavailability. *The American Journal of Clinical Nutrition*, *79(5)*, 727–747.

Mapari, S. A. S., Thrane, U., & Meyer, A. S. (2010). Fungal polyketide azaphilone pigments as future natural food colorants?. *Trends in Biotechnology*, *28(6)*, 300–307.

Marco, M. L., Heeney, D., Binda, S., Cifelli, C. J., Cotter, P. D., Foligné, B., Gänzle, M., Kort, R., Pasin, G., Pihlanto, A., Smid, E. J., & Hutkins, R. (2017). Health benefits of fermented foods: microbiota and beyond. *Current Opinion in Biotechnology, 44,* 94–102.

Marín, J., Zalacain, A., De Miguel, C., Alonso, G. L., & Salinas, M. R. (2005). Stir bar sorptive extraction for the determination of volatile compounds in oak-aged wines. *Journal of chromatography. A, 1098 (1-2),* 1–6.

Marquez, A., Serratosa, M. P., & Merida, J. (2013). Pyranoanthocyanin derived pigments in wine: structure and formation during winemaking. *Journal of Chemistry*, *2013*, 1–15.

Martillanes, S., Rocha-Pimienta, J., Cabrera-Bañegil, M., Martín-Vertedor, D., & Delgado-Adámez, J. (2017). Application of phenolic compounds for food preservation: Food additive and active packaging. In *Phenolic Compounds–Biological Activity.* IntechOpen, 39-58. London, UK:

Mateus, N., & de Freitas, V. (2001). Evolution and stability of anthocyanin-derived pigments during Port wine aging. *Journal of Agricultural and Food Chemistry, 49(11),* 5217–5222.

Mateus, N., Oliveira, J., Haettich-Motta, M., & de Freitas, V. (2004a). New family of bluish pyranoanthocyanins. *Journal of Biomedicine and Biotechnology, 2004(5)*, 299–305.

Mateus, N., Oliveira, J., Santos-Buelga, C., Silva, A. M.S., & de Freitas, V. (2004b). NMR structure characterization of a new vinylpyranoanthocyanin–catechin pigment (a portisin). *Tetrahedron Letters, 45(17),* 3455–3457.

Mateus, N., Silva, A. M. S., Rivas-Gonzalo, J. C., Santos-Buelga, C., & de Freitas, V. (2003). A new class of blue anthocyanin-derived pigments isolated from red wines. *Journal of Agricultural and Food Chemistry, 51(7)*, 1919–1923.

Mazza, G., & Brouillard, R. (1987). Recent developments in the stabilization of anthocyanins in food products. *Food Chemistry*, *25(3)*, 207–225.

Mazza, G., & Brouillard, R. (1990). The mechanism of co-pigmentation of anthocyanins in aqueous solutions. *Phytochemistry*, *29(4)*, 1097–1102.

Mazza, G., & Miniati, E. (1993). Anthocyanins in fruits, vegetables, and grains. Boca Raton: CRC Press.

McCann, D., Barrett, A., Cooper, A., Crumpler, D., Dalen, L., Grimshaw, K., Kitchin, E., Lok, K., Porteous, L., Prince, E., Sonuga-Barke, E., Warner, J. O., & Stevenson, J. (2007). Food additives and hyperactive behaviour in 3-year-old and 8/9-year-old children in the community: a randomised, double-blinded, placebo-controlled trial. *The Lancet, 370(9598)*, 1560–1567.

McDonnell, G., & Russell, A. D. (1999). Antiseptics and disinfectants: activity, action, and resistance. *Clinical Microbiology Reviews, 12(1),* 147-179.

McKone, H. T. (1991). The history of food colorants before aniline dyes. *Bulletin for the History of Chemistry, 10*, 25–31.

Méndez-Gallegos, S., Panzavolta, T., & Tiberi, R. (2003). Carmine cochineal *Dactylopius coccus* costa (Rhynchota: Dactylopiidae): significance, production and use. *Advances in Horticultural Science, 17(3)*, 165–171.

Mercadante, A. Z., & Bobbio, F. O. (2008). Anthocyanins in foods: occurrence and physicochemical properties. In *Food colorants: Chemical and Functional Properties*, Sosaciu, C., Ed.; CRC Press: Boca Raton,FL., 241–276.

Middelhoven, W. J., and M. D. S. Gelpke. (1995). Partial conversion of cinnamic acid into styrene by growing cultures and cell-free extracts of the yeast *Cryptococcus elinovii. Antonie Leeuwenhoek, 67*, 217–219.

Molenaar, D., Bosscher, J. S., Brink, B. ten, Driessen, A. J., & Konings, W. N. (1993). Generation of a proton motive force by histidine decarboxylation and electrogenic histidine/histamine antiport in *Lactobacillus buchneri. Journal of Bacteriology, 175(10),* 2864–2870.

Mollov, P., Mihalev, K., Shikov, V., Yoncheva, N., & Karagyozov, V. (2007). Colour stability improvement of strawberry beverage by fortification with polyphenolic copigments naturally occurring in rose petals. *Innovative Food Science and Emerging Technologies, 8(3),* 318–321.

Monagas, M., Martín-Álvarez, P. J., Gómez-Cordovés, C., & Bartolomé, B. (2007). Effect of the modifier (Graciano vs. Cabernet sauvignon) on blends of Tempranillo wine during ageing in the bottle. II. Colour and overall appreciation. *LWT - Food Science and Technology, 40(1),* 107–115.

Montilla, E. C., Arzaba, M. R., Hillebrand, S., & Winterhalter, P. (2011). Anthocyanin composition of black carrot (*Daucus carota* ssp. sativus var. *atrorubens* Alef.) cultivars Antonina, Beta Sweet, Deep Purple, and Purple Haze. *Journal of Agricultural and Food Chemistry, 59(7)*, 3385–3390.

Morais, H., Ramos, C., Forgásc, E., Cserháti, T., Matos, N., Almeida, V., & Oliveira, J. (2002). Stability of anthocyanins extracted from grape skins. *Chromatographia, 56(1),* 173-175.

Morata, A., Benito, S., Loira, I., Palomero, F., González, M. C., & Suárez-Lepe, J. A. (2012). Formation of pyranoanthocyanins by *Schizosaccharomyces pombe* during the fermentation of red must. *International Journal of Food Microbiology, 159(1)*, 47–53.

Morata, A., Gómez-Cordovés, M. C., Calderón, F., & Suárez, J. A. (2006). Effects of pH, temperature and $SO_2$ on the formation of pyranoanthocyanins during red wine fermentation with two species of *Saccharomyces*. *International journal of Food Microbiology, 106(2)*, 123–129.

Morata, A., Gómez-Cordovés, M. C., Suberviola, J., Bartolomé, B., Colomo, B., & Suárez, J. A. (2003). Adsorption of anthocyanins by yeast cell walls during the fermentation of red wines. *Journal of Agricultural and Food Chemistry, 51(14),* 4084–4088.

Mori, K., Sugaya, S., & Gemma, H. (2005). Decreased anthocyanin biosynthesis in grape berries grown under elevated night temperature condition. *Scientia Horticulturae, 105(3)*, 319–330.

Mortensen, A. (2006). Carotenoids and other pigments as natural colorants. *Pure and Applied Chemistry, 78(8),*1477-1491.

Motilva, M.J., Serra, A., & Macià, A. (2013). Analysis of food polyphenols by ultra high-performance liquid chromatography coupled to mass spectrometry: an overview. *Journal of Chromatography. A, 1292*, 66–82.

Nagel, C. W., & Wulf, L. W. (1979). Changes in the anthocyanins, flavonoids and hydroxycinnamic acid esters during fermentation and aging of Merlot and Cabernet Sauvignon. *American Journal of Enology and Viticulture, 30(2),* 111–116.

Ochoa, M. R., Kesseler, A. G., Vullioud, M. B., & Lozano, J. E. (1999). Physical and chemical characteristics of raspberry pulp: storage effect on composition and color. *LWT - Food Science and Technology, 32(3),* 149–153.

Oliveira, J., Azevedo, J., Silva, A. M. S., Teixeira, N., Cruz, L., Mateus, N., & de Freitas, V. (2010). Pyranoanthocyanin dimers: a new family of turquoise blue anthocyanin-derived pigments found in Port wine. *Journal of Agricultural and Food Chemistry, 58(8),* 5154–5159.

Oliveira, J., Mateus, N., & de Freitas, V. (2013). Network of carboxypyranomalvidin-3-O-glucoside (vitisin A) equilibrium forms in aqueous solution. *Tetrahedron Letters, 54(37),* 5106–5110.

Olthof, M. R., Hollman, P. C., & Katan, M. B. (2001). Chlorogenic acid and caffeic acid are absorbed in humans. *The Journal of Nutrition, 131(1),* 66–71.

Pan, Y., Zhu, Z., Huang, Z., Wang, H., Liang, Y., Wang, K., Lei, Q., & Liang, M. (2009). Characterisation and free radical scavenging activities of novel red pigment from *Osmanthus fragrans'* seeds. *Food Chemistry, 112(4),* 909–913.

Patras, A., Brunton, N. P., O'Donnell, C., & Tiwari, B. K. (2010). Effect of thermal processing on anthocyanin stability in foods; mechanisms and kinetics of degradation. *Trends in Food Science and Technology, 21(1),* 3–11.

Pisal, D. S., & Lele, S. S. (2005). Carotenoid production from microalga, *Dunaliella salina*. *Indian Journal of Biotechnology, 4,* 476–483.

Plumb, G. W., Garcia-Conesa, M. T., Kroon, P. A., Rhodes, M., Ridley, S., & Williamson, G. (1999). Metabolism of chlorogenic acid by human plasma, liver, intestine and gut microflora. *Journal of the Science of Food and Agriculture, 79(3),* 390–392.

Prior, R. L. (2003). Fruits and vegetables in the prevention of cellular oxidative damage. *The American Journal of Clinical Nutrition*, *78(3)*, 570-578.

Qi, W. W., Vannelli, T., Breinig, S., Ben-Bassat, A., Gatenby, A. A., Haynie, S. L., & Sariaslani, F. S. (2007). Functional expression of prokaryotic and eukaryotic genes in *Escherichia coli* for conversion of glucose to *p*-hydroxystyrene. *Metabolic Engineering, 9(3),* 268-276.

Quina, F. H., & Bastos, E. L. (2018). Chemistry inspired by the colors of fruits, flowers and wine. *Anais da Academia Brasileira de Ciencias, 90(1),* 681–695.

Reguant, C., Bordons, A., Arola, L., & Rozes, N. (2000). Influence of phenolic compounds on the physiology of *Oenococcus oeni* from wine. *Journal of Applied Microbiology, 88(6),* 1065-1071.

Rein, M. J., & Heinonen, M. (2004). Stability and enhancement of berry juice color. *Journal of Agricultural and Food Chemistry, 52(10),* 3106-3114.

Rein, M. J., Ollilainen, V., Vahermo, M., Yli-Kauhaluoma, J., & Heinonen, M. (2005). Identification of novel pyranoanthocyanins in berry juices. *European Food Research and Technology*, *220(3-4)*, 239–244.

Rentzsch, M., Schwarz, M., & Winterhalter, P. (2007). Pyranoanthocyanins – an overview on structures, occurrence, and pathways of formation. *Trends in Food Science and Technology, 18(10),* 526–534.

Rentzsch, M., Weber, F., Durner, D., Fischer, U., & Winterhalter, P. (2009). Variation of pyranoanthocyanins in red wines of different varieties and vintages and the impact of pinotin A addition on their color parameters. *European Food Research and Technology*, *229(4)*, 689–696.

Ribéreau-Gayon, P., Dubourdieu, D., Donèche, B., & Lonvaud, A. (2006). The microbiology of wine and vinifications. Handbookof Enology. John Wiley & Sons Ltd, Chichester, UK

Rice-Evans, C. A., Miller, N. J., & Paganga, G. (1996). Structure-antioxidant activity relationships of flavonoids and phenolic acids. *Free Radical Biology and Medicine, 20(7),* 933–956.

Richard, P., Viljanen, K., & Penttilä, M. (2015). Overexpression of *PAD1* and *FDC1* results in significant cinnamic acid decarboxylase activity in *Saccharomyces cerevisiae*. *AMB Express, 5(1),* 12.

Rodriguez, A., Martnez, J. A., Flores, N., Escalante, A., Gosset, G., & Bolivar, F. (2014). Engineering *Escherichia coli* to overproduce aromatic amino acids and derived compounds. *Microbial Cell Factories, 13(1),* 126.

Rodriguez-Saona, L. E., Giusti, M. M., & Wrolstad, R. E. (1999). Color and pigment stability of red radish and red-fleshed potato anthocyanins in juice model systems. *Journal of Food Science*, *64(3)*, 451–456.

Romero, C., & Bakker, J. (1999). Interactions between grape anthocyanins and pyruvic acid, with effect of pH and acid concentration on anthocyanin composition and color in model solutions. *Journal of Agricultural and Food Chemistry, 47(8),* 3130-3139.

Romero, C., & Bakker, J. (2000). Effect of storage temperature and pyruvate on kinetics of anthocyanin degradation, vitisin A derivative formation, and color characteristics of model solutions. *Journal of Agricultural and Food Chemistry, 48(6),* 2135-2141.

Rozès, N., & Peres, C. (1998). Effects of phenolic compounds on the growth and the fatty acid composition of *Lactobacillus plantarum*. *Applied Microbiology and Biotechnology*, *49(1)*, 108–111.

Sadilova, E., Stintzing, F. C., & Carle, R. (2006). Thermal degradation of acylated and nonacylated anthocyanins. *Journal of Food Science, 71(8),* C504-C512.

Saltmarsh, M. (2014). Recent trends in the use of food additives in the United Kingdom. *Journal of the Science of Food and Agriculture, 95(4),* 649–652.

Sánchez-Maldonado, A. F., Schieber, A., & Gänzle, M. G. (2011). Structure-function relationships of the antibacterial activity of phenolic acids and their metabolism by lactic acid bacteria. *Journal of Applied Microbiology*, *111(5)*, 1176–1184.

Sariaslani, F. S. (2007). Development of a combined biological and chemical process for production of industrial aromatics from renewable resources. *Annual Review of Microbiology, 61,* 51-69.

Sarni-Manchado, P., Fulcrand, H., Souquet, J. M., Cheynier, V., & Moutounet, M. (1996). Stability and color of unreported wine anthocyanin-derived pigments. *Journal of Food Science, 61(5),* 938–941.

Sasaki, Y. F., Kawaguchi, S., Kamaya, A., Ohshita, M., Kabasawa, K., Iwama, K., Taniguchi, K., & Tsuda, S. (2002). The comet assay with 8 mouse organs: results with 39 currently used food additives. *Mutation Research/Genetic Toxicology and Environmental Mutagenesis, 519(1-2),* 103–119.

Sawa, T., Nakao, M., Akaike, T., Ono, K., & Maeda, H. (1999). Alkylperoxyl radical-scavenging activity of various flavonoids and other phenolic compounds: implications for the anti-tumor-promoter effect of vegetables. *Journal of Agricultural and Food Chemistry, 47(2)*, 397–402.

Scalbert, A., & Williamson, G. (2000). Dietary intake and bioavailability of polyphenols. *The Journal of Nutrition*, *130(8)*, 2073-2085.

Schab, D. W., & Trinh, N.-H. T. (2004). Do artificial food colors promote hyperactivity in children with hyperactive syndromes? A meta-analysis of double-blind placebo-controlled trials. *Journal of Developmental and Behavioral Pediatrics, 25(6),* 423–434.

Schieber, A., Stintzing, F. C., & Carle, R. (2001). By-products of plant food processing as a source of functional compounds — recent developments. *Trends in Food Science and Technology*, *12(11)*, 401–413.

Schieberle, P. (1991). Primary odorants in popcorn. *Journal of Agricultural and Food Chemistry, 39(6),* 1141–1144.

Schütz, K., Kammerer, D., Carle, R., & Schieber, A. (2004). Identification and quantification of caffeoylquinic acids and flavonoids from artichoke (*Cynara scolymus* L.) heads, juice, and pomace by HPLC-DAD-

ESI/$MS^n$. *Journal of Agricultural and Food Chemistry, 52(13),* 4090-4096.

Schwarz, M., & Winterhalter, P. (2003). A novel synthetic route to substituted pyranoanthocyanins with unique colour properties. *Tetrahedron Letters, 44(41),* 7583–7587.

Schwarz, M., Jerz, G., & Winterhalter, P. (2003a). Isolation and structure of Pinotin A, a new anthocyanin derivate from Pinotage wine. *Vitis: Journal of Grapevine Research*, *42(2)*, 105–106.

Schwarz, M., Quast, P., Baer, D. von, & Winterhalter, P. (2003b). Vitisin A content in chilean wines from *Vitis vinifera* Cv. Cabernet Sauvignon and contribution to the color of aged red wines. *Journal of Agricultural and Food Chemistry, 51(21),* 6261–6267.

Schwarz, M., Wabnitz, T. C., & Winterhalter, P. (2003c). Pathway leading to the formation of anthocyanin-vinylphenol adducts and related pigments in red wines. *Journal of Agricultural and Food Chemistry, 51(12),* 3682–3687.

Schwarz, M., Wray, V., & Winterhalter, P. (2004). Isolation and identification of novel pyranoanthocyanins from black carrot (*Daucus carota* L.) juice. *Journal of Agricultural and Food Chemistry, 52(16),* 5095–5101.

Septembre-Malaterre, A., Remize, F., & Poucheret, P. (2017). Fruits and vegetables, as a source of nutritional compounds and phytochemicals: Changes in bioactive compounds during lactic fermentation. *Food Research International*, *104*, 86-99.

Sharma, V., McKone, H. T., & Markow, P. G. (2011). A global perspective on the history, use, and identification of synthetic food dyes. *Journal of Chemical Education, 88(1),* 24–28.

Shimada, K., Kimura, E., Yasui, Y., Tanaka, H., Matsushita, S., Hagihara, H. Nagakura, M., & Kawahisa, M. (1992). Styrene formation by the decomposition by *Pichia carsonii* of *trans*-cinnamic acid added to a ground fish product. *Applied Environmental Microbiology, 58(5),* 1577-1582.

Shipp, J., & Abdel-Aal, E. S. M. (2010). Food applications and physiological effects of anthocyanins as functional food ingredients. *The Open Food Science Journal, 4(1),* 7–22.

Shukla, D., & Vankar, P. S. (2013). Natural dyeing with black carrot: new source for newer shades on silk. *Journal of Natural Fibers, 10(3),* 207-218.

Sigurdson, G. T., Tang, P., & Giusti, M. M. (2017). Natural colorants: food colorants from natural sources. *Annual Review of Food Science and Technology, 8(1),* 261–280.

Silva, I., Campos, F. M., Hogg, T., & Couto, J. A. (2011). Factors influencing the production of volatile phenols by wine lactic acid bacteria. *International Journal of Food Microbiology, 145(2-3),* 471–475.

Singh, M. P., Petersen, P. J., Weiss, W. J., Kong, F., & Greenstein, M. (2000). Saccharomicins, novel heptadecaglycoside antibiotics produced by *Saccharothrix espanaensis*: antibacterial and mechanistic activities. *Antimicrobial agents and chemotherapy, 44(8),* 2154-2159.

Soares, A. D. S. L. F. (2014). *Characterization of phenolic acid reductase and decarboxylase activities of lactic acid bateria* (Doctoral dissertation). University of Porto.

Socaciu, C. (2008). Food colorants, chemical and functional properties. Boca Raton, Florida, USA: CRC Press

Solymosi, K., Latruffe, N., Morant-Manceau, A., & Schoefs, B. (2015). Food colour additives of natural origin. *Colour Additives for Foods and Beverages*, 3–34.

Sousa, A., Cabrita, L., Araújo, P., Mateus, N., Pina, F., & de Freitas, V. (2014). Color stability and spectroscopic properties of deoxyvitisins in aqueous solution. *New Journal Chemistry*, *38(2)*, 539–544.

Sousa, M. M., Melo, M. J., Parola, A. J., Morris, P. J. T., Rzepa, H. S., & Seixas de Melo, J. S. (2008). A study in mauve: unveiling Perkin's dye in historic samples. *Chemistry-A European Journal, 14(28),* 8507–8513.

Spano, G., & Massa, S. (2006). Environmental stress response in wine lactic acid bacteria: beyond *Bacillus subtilis*. *Critical Reviews in Microbiology, 32(2),* 77-86.

Spinler, J. K., Taweechotipatr, M., Rognerud, C. L., Ou, C. N., Tumwasorn, S., & Versalovic, J. (2008). Human-derived probiotic *Lactobacillus reuteri* demonstrate antimicrobial activities targeting diverse enteric bacterial pathogens. *Anaerobe, 14(3),* 166-171.

Spolaore, P., Joannis-Cassan, C., Duran, E., & Isambert, A. (2006). Commercial applications of microalgae. *Journal of Bioscience and Bioengineering, 101(2)*, 87–96.

Steinkraus, K. H. (1997). Classification of fermented foods: worldwide review of household fermentation techniques. *Food Control, 8(5-6),* 311–317.

Stintzing, F. C., Stintzing, A. S., Carle, R., Frei, B., & Wrolstad, R. E. (2002). Color and antioxidant properties of cyanidin-based anthocyanin pigments. *Journal of Agricultural and Food Chemistry,50(21),* 6172–6181.

Stintzing, F. C., & Carle, R. (2004). Functional properties of anthocyanins and betalains in plants, food, and in human nutrition. *Trends in Food Science and Technology, 15(1),* 19–38.

Stolarcyk, J., & Janick, J. (2011). Carrot: history and iconography. *Chronica Horticulturae*, *51*, 13–18.

Sun, C., Zheng, Y., Chen, Q., Tang, X., Jiang, M., Zhang, J., Li, X., & Chen, K. (2012). Purification and anti-tumour activity of cyanidin-3-O-glucoside from Chinese bayberry fruit. *Food Chemistry, 131(4),* 1287–1294.

Sun, L. H., Lv, S. W., Yu, F., Li, S. N., & He, L. Y. (2018). Biosynthesis of 4-vinylguaiacol from crude ferulic acid by *Bacillus licheniformis* DLF-17056. *Journal of Biotechnology*, *281*, 144-149.

Svensson, L., Sekwati-Monang B., Lutz, D. L., Schieber, A., & Gänzle, M. G. (2010). Phenolic acids and flavonoids in nonfermented and fermented red sorghum (*Sorghum bicolor* (L.) Moench). *Journal of Agricultural and Food Chemistry, 58(16),* 9214-9220.

Szwajgier, D., & Jakubczyk, A. (2011). Production of extracellular ferulic acid esterases by *Lactobacillus* strains using natural and synthetic carbon sources. *Acta Scientiarum Polonorum, Technologia Alimentaria 10(3),* 287-302.

Takaichi, A., Okamoto, T., Yamada, T., Hatai, R., & Sato, T. (2003). *Drinks containing Cochineal colorant and method of preventing discoloration thereof.* U.S. Patent No. 6,630,186. Washington, DC: USA.

Talcott, S. T., Brenes, C. H., Pires, D. M., & Del Pozo-Insfran, D. (2003). Phytochemical stability and color retention of copigmented and processed muscadine grape juice. *Journal of Agricultural and Food Chemistry, 51(4),* 957-963.

Taofiq, O., González-Paramás, A. M., Barreiro, M. F., & Ferreira, I. C. F. R. (2017). Hydroxycinnamic acids and their derivatives: cosmeceutical significance, challenges and future perspectives, a review. *Molecules, 22(2), 281*

Tarozzi, A., Morroni, F., Hrelia, S., Angeloni, C., Marchesi, A., Cantelli-Forti, G., & Hrelia, P. (2007). Neuroprotective effects of anthocyanins and their in vivo metabolites in SH-SY5Y cells. *Neuroscience Letters, 424(1),* 36–40.

Teixeira, J., Gaspar, A., Garrido, E. M., Garrido, J., & Borges, F. (2013). Hydroxycinnamic acid antioxidants: an electrochemical overview. *BioMed Research International*, *2013*, Article ID 251754, 11.

Timberlake, C. F., & Henry, B. S. (1986). Plant pigments as natural food colours. *Endeavour*, *10(1)*, 31–36.

Toba, T., Samant, S. K., Yoshioka, E., & Itoh, T. (1991). Reutericin 6, a new bacteriocin produced by *Lactobacillus reuteri* LA 6. *Letters in Applied Microbiology, 13(6),* 281-286.

Toci, A. T., & Farah, A. (2008). Volatile compounds as potential defective coffee beans' markers. *Food Chemistry, 108(3),* 1133–1141.

Tomaro-Duchesneau, C., Saha, S., Malhotra, M., Coussa-Charley, M., Al-Salami, H., Jones, M., Labbe, A., & Prakash, S. (2012). *Lactobacillus fermentum* NCIMB 5221 has a greater ferulic acid production compared to other ferulic acid esterase producing Lactobacilli. *International Journal of Probiotics and Prebiotics, 7(1),* 23-32.

Topakas, E., Vafiadi, C., & Christakopoulos, P. (2007). Microbial production, characterization and applications of feruloyl esterases. *Process Biochemistry, 42(4)*, 497-509.

Torres-Mancera, M. T., Cordova-López, J., Rodríguez-Serrano, G., Roussos, S., Ramírez-Coronel, M. A., Favela-Torres, E., & Saucedo-Castañeda, G. (2011). Enzymatic extraction of hydroxycinnamic acids from coffee pulp. *Food Technology and Biotechnology, 49(3),* 369-373.

Trouillas, P., Sancho-García, J. C., de Freitas, V., Gierschner, J., Otyepka, M., & Dangles, O. (2016). Stabilizing and modulating color by copigmentation: insights from theory and experiment. *Chemical Reviews, 116(9)*, 4937–4982.

Türker, N., & Erdoğdu, F. (2006). Effects of pH and temperature of extraction medium on effective diffusion coefficient of anthocynanin pigments of black carrot (*Daucus carota* var. L.). *Journal of Food Engineering, 76(4),* 579-583.

Türkyilmaz, M., & Özkan, M. (2012). Kinetics of anthocyanin degradation and polymeric colour formation in black carrot juice concentrates during storage. *International Journal of Food Science and Technology, 47(11)*, 2273-2281.

Tymczyszyn, E. E., Gómez-Zavaglia, A., & Disalvo, E. A. (2005). Influence of the growth at high osmolality on the lipid composition, water permeability and osmotic response of *Lactobacillus bulgaricus*. *Archives of Biochemistry and Biophysics, 443(1-2)*, 66–73.

Ulmer, H. M., Herberhold, H., Fahsel, S., Gänzle, M. G., Winter, R., & Vogel, R. F. (2002). Effects of pressure-induced membrane phase transitions on inactivation of HorA, an ATP-dependent multidrug resistance transporter, in *Lactobacillus plantarum*. *Applied and Environmental Microbiology, 68(3)*, 1088–1095.

Vallverdú-Queralt, A., Biler, M., Meudec, E., Le Guernevé, C., Vernhet, A., Mazauric, J. P., Legras, J.L., Loonis, M., Trouillas, P., Cheynier, V., & Dangles, O. (2016). *p*-Hydroxyphenyl-pyranoanthocyanins: an experimental and theoretical investigation of their acid-base properties and molecular interactions. *International Journal of Molecular Sciences, 17(11)*, 1842.

Van Beek, S., & Priest, F. G. (2000). Decarboxylation of substituted cinnamic acids by lactic acid bacteria isolated during malt whisky fermentation. *Applied Environmental Microbiology, 66(12)*, 5322-5328.

Vannelli, T., Qi, W. W., Sweigard, J., Gatenby, A. A., & Sariaslani, F. S. (2007). Production of *p*-hydroxycinnamic acid from glucose in *Saccharomyces cerevisiae* and *Escherichia coli* by expression of heterologous genes from plants and fungi. *Metabolic Engineering, 9(2)*, 142-151.

Vargas-Tah, A., Martínez, L. M., Hernández-Chávez, G., Rocha, M., Martínez, A., Bolívar, F., & Gosset, G. (2015). Production of cinnamic and *p*-hydroxycinnamic acid from sugar mixtures with engineered *Escherichia coli*. *Microbial Cell Factories, 14(1)*, 6.

Velmurugan, P., TamilSelvi, A., Lakshmanaperumalsamy, P., Park, J., & Oh, B.T. (2013). The use of cochineal and *Monascus purpureus* as dyes for cotton fabric. *Coloration Technology, 129(4)*, 246–251.

Venil, C. K., Zakaria, Z. A., & Ahmad, W. A. (2013). Bacterial pigments and their applications. *Process Biochemistry, 48(7)*, 1065–1079.

Vivar-Quintana, A. M., Santos-Buelga, C., Francia-Aricha, E., & Rivas-Gonzalo, J. C. (1999). Formation of anthocyanin-derived pigments in experimental red wines / Formación de pigmentos derivados de antocianos en vinos tintos experimentales. *Food Science and Technology International, 5(4)*, 347–352.

Vivar-Quintana, A. M., Santos-Buelga, C., & Rivas-Gonzalo, J. C. (2002). Anthocyanin-derived pigments and colour of red wines. *Analytica Chimica Acta*, *458(1)*, 147–155.

Wang, H., Pan, Y., Tang, X., & Huang, Z. (2006). Isolation and characterization of melanin from *Osmanthus fragrans'* seeds. *LWT - Food Science and Technology, 39(5),* 496–502.

Wang, H., Race, E. J., & Shrikhande, A. J. (2003). Anthocyanin transformation in Cabernet Sauvignon wine during aging. *Journal of Agricultural and Food Chemistry*, *51(27)*, 7989–7994.

Wegkamp, A., Teusink, B., de Vos, W. M., & Smid, E. J. (2010). Development of a minimal growth medium for *Lactobacillus plantarum*. *Letters in Applied Microbiology, 50(1),* 57-64.

Williamson, G., Kroon, P. A., & Faulds, C. B. (1998). Hairy plant polysaccharides: a close shave with microbial esterases. *Microbiology, 144 (8),* 2011–2023.

Wissgott, U., & Bortlik, K. (1996). Prospects for new natural food colorants. *Trends in Food Science and Technology*, *7(9)*, 298–302.

Wrolstad, R. E. (2004). Anthocyanin pigments-bioactivity and coloring properties. *Journal of Food Science*, *69(5)*, C419-C425.

Wrolstad, R. E., & Culver, C. A. (2012). Alternatives to those artificial FD&C food colorants. *Annual Review of Food Science and Technology*, *3(1)*, 59–77.

Wrolstad, R. E., Durst, R. W., & Lee, J. (2005). Tracking color and pigment changes in anthocyanin products. *Trends in Food Science and Technology*, *16(9)*, 423–428.

Watts, K. T., Mijts, B. N., Lee, P. C., Manning, A. J., & Schmidt-Dannert, C. (2006). Discovery of a substrate selectivity switch in tyrosine ammonia-lyase, a member of the aromatic amino acid lyase family. *Chemistry and Biology, 13(12),* 1317-1326.

Wu, X., & Prior, R. L. (2005). Systematic identification and characterization of anthocyanins by HPLC-ESI-MS/MS in common foods in the United

States: fruits and berries. *Journal of Agricultural and Food Chemistry, 53(7)*, 2589–2599.

Xue, Z., McCluskey, M., Cantera, K., Sariaslani, F. S., & Huang, L. (2007a). Identification, characterization and functional expression of a tyrosine ammonia-lyase and its mutants from the photosynthetic bacterium *Rhodobacter sphaeroides*. *Journal of Industrial Microbiology and Biotechnology, 34(9),* 599-604.

Xue, Z., McCluskey, M., Cantera, K., Ben-Bassat, A., Sariaslani, F. S., & Huang, L. (2007b). Improved production of *p*-hydroxycinnamic acid from tyrosine using a novel thermostable phenylalanine/tyrosine ammonia lyase enzyme. *Enzyme and Microbial Technology, 42(1),* 58-64.

Yao, N., Lan, F., He, R.R., & Kurihara, H. (2010). Protective effects of bilberry (*Vaccinium myrtillus* L.) extract against endotoxin-induced uveitis in mice. *Journal of Agricultural and Food Chemistry, 58(8)*, 4731–4736.

Yılmaz, F. M., & Ersus Bilek, S. (2017). Natural colorant enrichment of apple tissue with black carrot concentrate using vacuum impregnation. *International Journal of Food Science and Technology, 52(6),* 1508-1516.

Zaldivar, J., & Ingram, L. O. (1999). Effect of organic acids on the growth and fermentation of ethanologenic *Escherichia coli* LY01. *Biotechnology and Bioengineering, 66(4),* 203-210.

Zhang, W., Seki, M., & Furusaki, S. (1997). Effect of temperature and its shift on growth and anthocyanin production in suspension cultures of strawberry cells. *Plant Science, 127(2)*, 207–214.

Zhang, L., Zhou, J., Liu, H., Khan, M. A., Huang, K., & Gu, Z. (2012). Compositions of anthocyanins in blackberry juice and their thermal degradation in relation to antioxidant activity. *European Food Research and Technology, 235(4),* 637-645.

Zheng, Y., Wang, S. Y., Wang, C. Y., & Zheng, W. (2007). Changes in strawberry phenolics, anthocyanins, and antioxidant capacity in response to high oxygen treatments. *LWT-Food Science and Technology, 40(1),* 49-57.

Zhong, J.J., & Yoshida, T. (1993). Effects of temperature on cell growth and anthocyanin production in suspension cultures of *Perilla frutescens*. *Journal of Fermentation and Bioengineering*, *76(6)*, 530–531.

www.ingramcontent.com/pod-product-compliance
Ingram Content Group UK Ltd.
Pitfield, Milton Keynes, MK11 3LW, UK
UKHW022001190726
13853UKWH00004B/1656

9 783736 975538